Globalisation and the Third World

Globalisation has been described as *the* major issue in the social sciences today. There is, however, little clarity about what it means and how it manifests itself in particular areas of concern.

Globalisation and the Third World examines these and other problems by bridging the gap between theoretical and empirical research and examining the processes of globalisation with reference to topical case studies and examples. Looking at the changing position of the developing world within the world system, each chapter focuses on particular issues which cut across communities, nations, regions and, in consequence, the world.

Exploring the manifestations of globalising processes in the world today, the authors examine aspects of development, migration, health, capital flows, the environment, popular culture in Latin America, religion and cultural imperialism and popular music. Focusing on the contradictory and hierarchical nature of processes of globalisation, the authors stress not only the effects of globalisation but also its limits, indicating that the process does not necessarily entail uniformity or harmony. While a 'global era' may mean the end of the 'Third World', it simultaneously means the intensification of uneven development.

Written in a lively and engaging style, *Globalisation and the Third World* offers both an invaluable introduction to a wide variety of development issues, as well as providing an important contribution to current debates around the issue of globalisation.

Ray Kiely and **Phil Marfleet** are both Senior Lecturers in the Department of Cultural Studies, University of East London.

Globalisation and the Third World

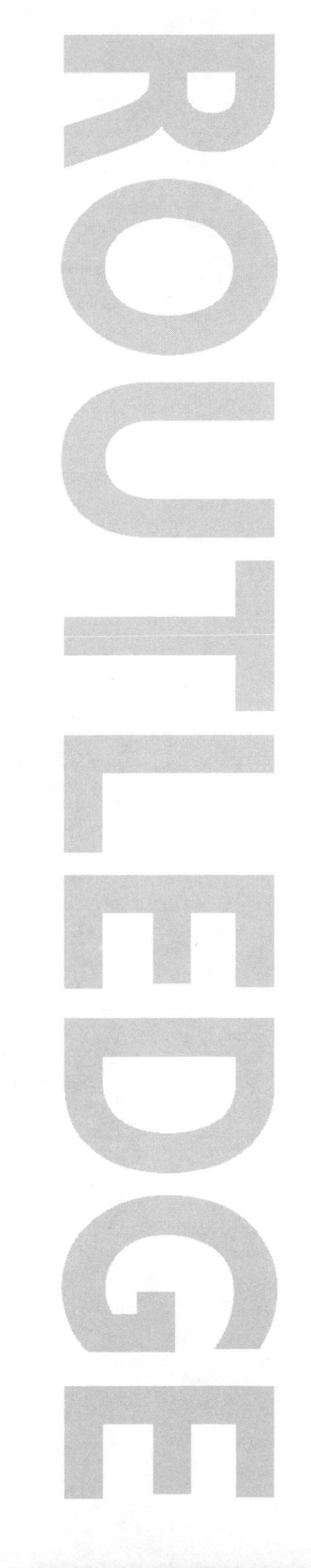

Edited by Ray Kiely
and
Phil Marfleet

LONDON AND NEW YORK

First published 1998
by Routledge
11 New Fetter Lane, London EC4P 4EE

Simultaneously published in the USA and Canada
by Routledge
29 West 35th Street, New York, NY 10001

Reprinted 2000

Routledge is an imprint of the Taylor & Francis Group

Typeset in Garamond 3 by Keystroke, Jacaranda Lodge, Wolverhampton
Printed and bound in Great Britain by TJ International Ltd, Padstow, Cornwall

British Library Cataloguing in Publication Data
A catalogue record for this book is available from the British Library

Library of Congress Cataloguing in Publication Data
Globalisation and the Third World / edited by Ray Kiely and Phil Marfleet.
p. cm.
Includes bibliographical references and index.
1. Economic development. 2. Developing countries. I. Kiely, Ray, 1948– . II. Marfleet, Phil, 1962– .
HD82.G553 1998
337′09172′4—dc21 97–48955
CIP
AC

ISBN 0–415–14076–5 (hbk)
ISBN 0–415–14077–3 (pbk)

Contents

Contributors

Suzanne Biggs is Senior Lecturer in the Department of Cultural Studies at the University of East London.

David Hesmondhalgh is Lecturer in Media and Communications Studies at Goldsmiths' College, University of London. He is co-editor (with Georgina Born) of *Western Music and its 'Others'* (1998) and is currently writing a book on *The Cultural Politics of Music.*

Ray Kiely is Senior Lecturer in the Department of Cultural Studies at the University of East London. His most recent book is *Industrialisation and Development: a Comparative Analysis* (1997).

Maureen Larkin is Senior Lecturer in Health Studies at the University of East London.

Phil Marfleet is Senior Lecturer in the Department of Cultural Studies at the University of East London, where he co-ordinates the MA programme in Refugee Studies.

Vivian Schelling is Senior Lecturer in the Department of Cultural Studies at the University of East London, and co-author (with William Rowe) of *Memory and Modernity: Popular Culture in Latin America* (1991).

Globalisation, (post-)modernity and the Third World

Ray Kiely

introduction

The 1990s have seen a boom in writing about globalisation. According to one sociologist (Waters 1995: 1), it is *the* concept of the 1990s, 'a key idea by which we understand the transition of human society into the third millennium'. This book is a contribution to the debates over what is a difficult concept, and it takes a different approach to most 'mainstream' globalisation theory. Much of the debate surrounding 'globalisation' has been extremely abstract. There is often a lack of clarity in definitions of the term, its novelty and how it is experienced by people throughout the world. The chapters that follow attempt to address these problems by concretising the concept, by bringing it down to earth. This position is not to endorse an atheoretical populism – on the contrary all of the chapters are informed by theory. On the other hand, it is to endorse a position whereby theory and evidence are brought into a closer relationship, one which makes theory more grounded.[1] The book therefore examines concrete issues – capital flows, migration, health, the environment, popular culture, the media industries and religion – and how these have become, or are becoming, more 'globalised'.

The chapters also address the issue of power and in particular the inequalities that continue to exist in the global order. Despite the proliferation of material on globalisation in recent years, few writers have addressed the questions of the so-called Third World and of development.[2] Indeed, some of the most optimistic accounts of globalisation (particularly of the economy) dispense with analysis of relations of power altogether.[3]

The problems of globalisation, power and the Third World are addressed in each of the chapters that follow. Before moving on, however, some more general comment about these concerns is required. The remainder of this introduction examines three issues: globalisation, development and the Third World, and global inequalities and their specific manifestations. The first section, on globalisation, discusses two related approaches to globalisation – from social theory and mass communications research. In assessing the main claims of these approaches, I argue that their approaches are suggestive, but also underestimate the limitations of globalising processes and – in so far as these have occurred – the degree to which these same processes are differentially experienced throughout the world. The second section, on development and the Third World, focuses on one ostensibly 'global' account of the world from the tradition of radical underdevelopment theory. In rejecting the claims of this particular approach, I suggest that this theory lacks an adequate account of historical process.

In the third section (pp. 10–16), I draw on these critiques to present an alternative approach, which is taken up by each of the chapters in the rest of the book. Our approach is (contrary to the approaches discussed in the first section (pp. 3–6)) sensitive to the differential experience of 'actually existing globalisation', and at the same time (contrary to the approach discussed in the second section (pp. 6–10)) we examine *globalisation not as a new theory, but as a set of historical and social processes*, which themselves have some limits. The concern in this book is therefore not with globalisation as a grand theory, but with globalisation as an agenda for concrete empirical research. In this spirit the third section surveys some examples of globalising processes, providing an introduction to the cases taken up in later chapters.

Globalisation

Globalisation refers to a world in which societies, cultures, polities and economies have, in some sense, come closer together. According to Giddens (1990: 64), the concept can be defined as 'the intensification of worldwide social relations which link distant localities in such a way that local happenings are shaped by events occurring many miles away and vice versa'. Thus, the job of a coal miner in Britain may depend on events in South Africa or Poland as much as on local management or national government decisions.

But the concept does not just refer to global interconnectedness. Globalisation 'is best understood as expressing fundamental aspects of time–space distanciation. Globalisation concerns the intersection of presence and absence, the interlacing of social events and social relations "at a distance" with local contextualities' (Giddens 1991: 21). David Harvey (1989: 240), too, refers to the fact that globalisation describes our changing experience of time and space, or 'time–space compression'.

These definitions therefore refer to two key factors. First, more and more parts of the world are drawn into a global system and so are affected by what happens elsewhere. I return to this point below. Second, there is a sense that we know what is happening elsewhere more quickly, which in turn affects our sense of space and place. Thus, for example, television coverage of the Gulf War was live and focused on both sides (the US-dominated forces and the Iraqis) of the conflict. The war was conducted in the real war time zone in the Middle East, but also with attention paid to time in both Europe and the United States. Moreover, the coverage was effectively controlled by the US military so that viewers were presented with a sanitised war, in which casualties were supposedly kept to a minimum (Taylor 1992). The selective use of television reportage across the world, and the speed at which it was transmitted, meant that for many (though not all) people the war was a kind of hyper-real event (Walsh 1995).

When the term globalisation is used in this second sense – to refer to 'time–space compression' – there is a close link with the development of new communications technologies and the 'postmodern condition'.[4] The development of satellite television and information technology are two major factors in the development of a new information society. For some writers, however, this increase in images and representations does not lead to an *informed* society. On the contrary, '[i]nformation dissolves meaning and the social into a sort of nebulous state leading not at all to a surfeit of innovation but to the very contrary, to total entropy' (Baudrillard 1983: 100). For Baudrillard,[5] then, the new communications technologies have promoted a global culture of instant but meaningless communication, in which time and space horizons have collapsed. He argues that this new social order is based on the dominance of signs rather than material production – the 'metallurgic' society has become a 'semiurgic society' (Baudrillard 1981: 185). In this society the boundary between image and reality implodes, so that the latter becomes meaningless. There is no longer any space in which we can ground ourselves and our understanding of the world. History has vanished 'into a hyperspace where it loses all meaning' (Baudrillard 1988: 36). Given the collapse of time and space horizons, it would appear that geography has met the same fate (Meyrowitz 1986; O'Brien 1991).

While Baudrillard and other 'postmodern' writers point to important developments in the global order, he has stretched his case too far.[6] When he refers to the increased speed with which we are saturated with images, Baudrillard fails to state to whom this 'we' refers. Thus Waters (1995: 124), clearly influenced by Baudrillard, writes that 'material and power exchanges in the economic and political arenas are progressively becoming displaced by symbolic ones, that is by relationships based on values, preferences and tastes rather than by material inequality and constraint'. Given the figures on global poverty outlined in Chapter 1, this statement seems naive at best. Naivety can easily become offensive, especially when Baudrillard takes his notion of hyper-reality to the extreme position of denying that the Gulf War even occurred. It is one thing to recognise the difficulties that world viewers had in finding out the truth about the Gulf War, but for those thousands killed the war was all too real (Norris 1992).

What these comments suggest is that 'time–space' *compression* is not the same as time–space *destruction*, and that globalisation is experienced in different ways throughout the globe. As Morley and Robins point out:

> The particularity of place and culture can never be done away with, can never be absolutely transcended. Globalisation is, in fact, also associated with new dynamics of *re*-localisation. It is about the achievement of a new global–local nexus, about new and intricate relations between global space and local space.
>
> (Morley and Robins 1995: 116)

We may then be witnessing some important changes in the world, but these have not, and indeed cannot, abolish space. But we can go further than this, as Allen and Hamnett point out, for:

> the networks of social relationships stretched across space are not simply uneven in their global reach, they also work through *geographical difference and diversity*. Geography matters in this instance, precisely because global relations construct unevenness in their wake *and* operate through the pattern of uneven development laid down.
>
> (Allen and Hamnett 1995b: 235)

These observations are also relevant when one considers more optimistic scenarios of post-industrial futures. Much has been made of the potential of the Internet, as a basis for reconstructing and democratising social relations throughout the globe. Although it would be unwise to totally discount the ways that it can be used to rapidly disseminate progressive ideas across the globe,[7] there is still the reality of the massively unequal global distribution of communications. At least 80 per cent of the world's population still lack access to the most basic communications technologies, and nearly 50 countries have fewer than one telephone line per 100 people. There are more telephone lines in Manhattan than there are in the whole of sub-Saharan Africa. While the United States has 35 computers per 100 people, even rapidly developing South Korea has only 9, while for Ghana the figure is as low as 0.11. Although the number of Internet users has expanded dramatically in recent years, its use is still largely confined to Western Europe and the United States (*New Internationalist* 1996: 18–19). Moreover, the overwhelming proportion of Internet activity takes place at work (Castells 1996: 360–1). Even if computer prices fall to levels where they become easily affordable, there are other expenses for potential users. These include specialised cabling, advanced modems and online charges. Thus as Vincent points out:

> the market-place can be easily used to control access to the information superhighway. The greater the expense of access, the higher the likelihood that larger portions of the world's population – both developed as well as developing – will remain 'information poor'.
>
> (Vincent 1997: 187)

The over-optimism concerning the potential of the 'information society' is thus rooted in an old fashioned technological determinism, effectively satirised by Stallabrass:

> In this ostensibly democratic forum, a chairman of some Western conglomerate and an impoverished peasant in Central Africa will both use a device, much the size of a Walkman, to communicate by satellite with a panoply of open information systems.
>
> (Stallabrass 1995: 10)

Clearly, then, the information superhighway has passed by most of the world's population and is likely to do so for the foreseeable future. This point applies not only to the poorest parts of the so-called Third World, but also to many people living in global cities such as Los Angeles, where new telecommunications networks have produced 'electronic ghettoes', in which access is restricted to television screens. Clearly, then, time–space compression is experienced differently across the globe. For the people excluded, '[t]his is not even life in the slow lane. It is life on the hard shoulder' (Thrift 1995: 31).

The question of speed is therefore important but needs to be qualified. So too does the first sense in which globalisation was defined – that more and more people are drawn into the global system and so are affected by what happens elsewhere. For many parts of the world, and especially the Third World, this has been a fact of life for hundreds of years (Massey 1992). The experiences of colonial encounters, genocide and slavery, among others, show that the fate of most of Asia, Latin America, Africa and elsewhere has been closely tied to the expansion of Europe after 1492 (Stavrianos 1981).

The 'postmodern' turn in the social sciences in recent years is in part a recognition that modernist theories[8] at times either focused too narrowly on the national, or indeed were part of 'the technologies of colonial and imperialist governance' (Bhaba 1994: 195). Ironically, the postmodern turn is itself caught up in a similar problematic in that

> the Eurocentrically defined cultural conditions of a so-called post-modernity – irony, pastiche, the mixing of different histories, intertextuality, schizophrenia, cultural chasms, fragmentation, incoherence, disjunction of supposedly modern and pre-modern cultures – were characteristic of colonial societies, cultures, environments on the global periphery (in Calcutta, Hong Kong, Rio or Singapore) decades, if not centuries, before they appeared in Europe or the USA.
>
> (King 1995: 120–1; see also Morley 1996)

These points reinforce the argument already made. There is a need to look more closely at modernity 'in terms of space as well as time' (King 1995: 121). For our purposes, then, we need to look more closely at the questions of development and the Third World.

The Third World and development

The idea of development has a long history stretching back to the eighteenth century (Cowen and Shenton 1996). However, it was in the post-1945 period that it was consciously advocated as a way for 'the Rest' (or Third World) to become more like 'the West' (or First World). The context for this strategy of development was the beginning of the end of empire and the Cold War between capitalism and communism. The content of this strategy was economic growth through a partnership between the private and public sector, which was premised on the belief that the nation-state was a sovereign body. Mainstream theories of development, which took a number of forms between the 1940s and 1970s, were thus firmly centred on the primacy of the nation-state (Hettne 1995).

The post-1945 history of development is addressed in detail in Chapter 1. Here I want to focus less on these 'orthodox' approaches to development policy – which were Euro- or Americo-centric – and more on critical approaches to development. These are of interest because they represented a (partial) challenge to American hegemony (and with it Eurocentric discourses), and because their analyses went beyond the nation-state. In some respects then these were theories about globalisation which were critical of Western dominance of the world system.

The main critical theory I want to address is that of underdevelopment theory.[9] Although this has met with fierce (and deserved) criticism in recent years, it is of interest because of its status as an implicit theory of globalisation. The main advocate of this approach is Andre Gunder Frank. His basic thesis is that development and underdevelopment are two sides of the same coin – one leads to the other. The world has been capitalist since the sixteenth century, and in this world system,[10] metropoles develop by extracting the surplus (or underdeveloping) the satellites (Frank 1969: 9). For Frank, '[t]his capitalist system has at all times and in all places – as in its nature it must – produced both development and underdevelopment. The one is as much the product of the system, is just as "capitalist" as the other' (1969: 240).

For our immediate purpose, what is most interesting about this contention is that it avoids the binary oppositions of modern and traditional. Sixteenth-century Latin America – or indeed nineteenth-century Africa (Rodney 1974) – cannot be described as in any sense traditional because they are intrinsic parts of the modern, or capitalist, world. To his credit, then, Frank conceptualises modernity in terms of space as well as time. Development then cannot be a case of the Rest becoming more like the West precisely because the West is exploiting the Rest. For the Rest to really follow the West, it would literally have to colonise the latter – hardly an acceptable position for post-war American conservatives.

However, there are strong grounds for rejecting this theory as an adequate account of globalisation. Important linkages have existed between the West and the Rest, such as New World slavery, plunder in India and the scramble for Africa. These undoubtedly left parts of the world in a situation of *competitive disadvantage vis-à-vis* the colonial powers and contributed to some economic development in the metropolitan areas of the capitalist world (Bernstein *et al.* 1992). However, this does not mean that this exploitation was itself sufficient to promote industrialisation. Many of the early colonial powers such as Spain and Portugal remained predominantly feudal and agrarian in character, whereas the most successful early capitalist industrialiser, Britain, industrialised in the context of new capitalist social relations of production (Brenner 1977). In fact, trade *between* metropoles has historically been far greater than trade between core and periphery (De Vries 1976; O'Brien 1982). This suggests that the dominant powers in the global order have *marginalised* the rest of the world, a situation that continues – albeit in new forms – to this day (see Chapter 2).

Moreover, while on the one hand underdevelopment theory forces us to think about space – in that some regions develop at the expense of others – on the other, it conceptualises space in ways that mean that 'localities' have little capacity for meaningful activity. Local satellites or peripheries are homogenised as exploited areas to the extent that changes in them are explained solely by reference to the functional needs of the metropolitan areas (Booth 1985; Kiely 1995: ch. 3). In this 'puppet master' theory of imperialism (Bernstein and Nicholas 1983), the strings of the local satellites are pulled by the all-powerful Western metropoles. In this way space is effectively destroyed as it operates in a one-way relationship with more powerful areas. Thus, Wallerstein (1983: 34–5), for example, effectively denies the efficacy of meaningful political action in the Third World.

Frank and others (Amin *et al.* 1982) have proposed one strategy for satellites to overcome underdevelopment. This was a complete delinking from the world economy in order to promote genuine development. In other words, a totally meaningless local, doomed to perpetual stagnation, could be replaced by a completely autonomous local, blessed with inevitable development (Brenner 1977; Gulalp 1986).

This dualism of total domination/complete authenticity has been replicated in recent years in some versions of cultural theory. The Western world has often made the Orientalist assumption that difference actually constitutes backwardness (Hall 1992; Said 1978). Europe therefore represented its colonised Other as inherently backward and in need of civilisation. In these postmodern times of (ostensibly) allowing the Other to speak, the conflation of difference with backwardness has been rightly rejected. However, in its place we are often given a simple celebration of difference and Otherness. But, as Ahmad (1994: 90) points out, this Manichaean approach tends to reduce difference to something that is inherent, an 'epistemological category or perennial ontological condition'. At its worst this represents the return of the Noble Savage, a romantic (and patronising) celebration of the Other. This can lead to a blinkered celebration of Third World nationalisms,[11] at least if they are prepared to 'go all the way' and completely break (or delink) from the evil West. Hence underdevelopment theory's uncritical approach to the tyranny of the Khmer Rouge in Cambodia from 1975 to early 1979 (Amin *et al.* 1982).

Some versions of postcolonial theory therefore replicate the Western binary opposition of modernity and tradition, albeit this time uncritically celebrating rather than condemning the latter. This failure to break with colonising discourses is also apparent in underdevelopment theory. As I have argued elsewhere:

> 'Genuine' capitalist development is ruled out by this theory, but this can be done only by having an implicit 'model' of what constitutes 'normal' capitalist development. In this sense the theory is a 'mirror image' of evolutionary theories, where 'inevitable development' in the case of modernization is replaced by 'inevitable stagnation or distortion' in underdevelopment and dependency.
>
> (Kiely 1995: 52)

Delinking from the West is proposed as an economic strategy so that the Rest can in fact catch up, and be like the West. The goal is the same, the difference is over the means to achieve it.

These critical comments suggest that underdevelopment theory, and its closely related variants, do not constitute an adequate theory of globalisation. In fact, underdevelopment theory actually puts forward a static, ahistorical model in which core countries exploit peripheral ones. The last 500 years of the world system can be reduced to this simple mechanism – an ongoing North–South global divide. The periphery or Third World has no history outside of its exploitation by the metropolitan countries. But is this a satisfactory account of even the last 200 years of world history? Given the rise and subsequent (relative) fall of Britain, the decline and much later (partial) resurgence of Spain, Japan's emergence as an economic superpower, the power of the United States in the twentieth century, the rise and fall of communism, independence in the nineteenth century and industrialisation in the twentieth for much of Latin America, Brazil's emergence as a regional power in that continent, independence in Africa, the end of apartheid in South Africa, independence in the Indian sub-continent, the rise of the newly industrialising countries (NICs) in East and South-East Asia, among many others, the answer must be negative. And these examples were only taken from politico-economic history, the exclusive concern of underdevelopment theory. What this theory lacks then is a sense of *historical process*. In so far as change is explained, it is deemed to be part of the never-ending logic of an omnipresent world system. But such functionalism – which 'knows' the answers before events have even occurred – can never get to grips with the contingencies, unpredictabilities and contradictions of this system.

This can be seen in the case of the rise of the East Asian newly industrialising countries since the 1970s. The rise of South Korea, Taiwan and others was a major factor in the undermining of underdevelopment as a major radical theory of development, and has led many to conclude that we have witnessed the 'end of the Third World' (Harris 1986). Underdevelopment theory was replaced by more flexible theories of dependency, which argued that the rise of East Asia did not constitute 'proper' capitalist development because it remained dependent on the core countries (Hart-Landsberg 1979; Frobel *et al.* 1980). This assertion carries with it all kinds of methodological and empirical difficulties (Bernstein 1979; Kiely 1994), but it is the reference to 'proper' which really gives the game away. For this implies again that the West is the norm, and so by definition other countries are deviations from that norm (Phillips 1977). Once again, concrete historical processes are reduced to dogmatic *a priori* theories.

In so far as core–periphery relations continue to exist, these take qualitatively new forms rather than a simple division between First and Third World nations (Lash and Urry 1994: 28–30).[12] In this way, core–periphery relations should be seen as an *effect*, rather than a defining *cause*, of the world system[13] (Kiely 1998a). Castells has usefully described the new hierarchies:

> While its effects reach out to the whole planet, *its actual operation and structure concern only segments of economic structures, countries, and regions, in proportions that vary according to the particular position of a country or region in the international division of labour* . . . Thus, while the informational economy shapes the entire planet, and in this sense it is indeed global, most people in the planet do not work for or buy from the informational/global economy.
>
> (Castells 1996: 102–3, emphasis in original)

Global inequalities and uneven development

The implications of this argument need to be spelled out. The idea of globalisation is important in that it refers to an increase in interdependency across space and an increase in the speed with which this relationship may occur. Related to this, there is an increased *sense* that we live in an interdependent world. Robertson (1990: 26) points out that the idea of societies being organised on a national basis does not necessarily contradict the process of globalisation, and that 'the prevalence of the national society in the twentieth century is *an aspect of globalization*' (emphasis in original). There is thus a global awareness that we live in nation-states, which thus constitutes a 'particularization of universalism' (Robertson 1991). This has contradictory effects as global homogeneity coexists with plurality as the 'sense that the world is one place also brings nations closer together in cultural prestige competitions' (Featherstone 1990b: 10). Thus, Robertson (1991) also refers to a 'universalization of particularism'. I expand on these points in discussing religion below.

But we must also remain aware that global connections are far from new and that globalisation is experienced differently for different people. Radical global theories of development[14] force us to think about global inequalities and hierarchies, but in a way that is insensitive to history and social change.

What I think is needed then is a sense of *globalisation as process*. This is the approach taken by the chapters in the rest of the book. Rather than attempting to elaborate a grand theory of globalisation, or indeed postmodernity, the chapters take the more concrete approach of examining globalisation as it is (differentially) experienced in spaces throughout the world. The chapters are therefore sensitive to the interaction of the global–local, to the reality of inequality and power, to diversity, and indeed to the limitations of globalisation. Thus, the unifying theme of these chapters is the reality of *uneven development* in the global system.

The chapters can be placed broadly in two categories, although it should be stressed that some cut across these categories (and also that there is no assumption that one category determines another). To complete this introduction, then, some comments should be made about the globalisation of (i) political economy; (ii) culture.

Political economy

The globalisation of the world economy refers to the ability of capital to move freely across national boundaries (Ohmae 1991). Such 'hyper-mobility' undermines the ability of nation-states to direct so-called 'national economies' (Radice 1984). This movement is said to be particularly strong for finance capital, which does not face the constraints of fixed locational costs associated with industrial capital. Facilitated by the development of the new communications technologies, finance capital can move rapidly around the globe, in the process undermining among other things the ability of nation-states to control their national currency. This is undoubtedly the most globalised form of economic activity, and it is in this sphere that writers have referred to the 'end of geography' (O'Brien 1991). From 1980 to 1990, the volume of cross-border transactions in equities grew from $120 billion a year to $1.4 trillion, and international bank lending stock rose from $324 billion to $7.5 trillion (Hoogvelt 1997: 78–80).

However, even in the case of finance capital there are limits. First, the number of financial goods that are sold in highly integrated world markets is relatively limited. Even stock markets are not fully integrated because few companies have a sufficiently global reputation for trading in their stock to be active outside of their home market. Second, domestic savings and investment rates are highly correlated among the core countries, and so domestic investment is constrained by the former, rather than being easily financed from other countries' savings (Wade 1996: 73–4). Third, and most significant for our purposes, is the fact that many parts of the world simply are left off this map and are not part of these global networks. In terms of finance the real global cities are places like New York, London and Tokyo rather than Port-au-Prince, Harare and Calcutta. The Third World's share of international bank lending declined to about 11 per cent at the end of the debt-ridden 1980s, and this proportion was highly concentrated in the boom economies of East and South-East Asia (Hoogvelt 1997: 88). Third World economies are however (unequally) incorporated into this global network of financial relationships, as free capital movement has increased the potential (and legality) of capital flight on the part of 'national' financial elites in the Third World. The impact of global finance has therefore *deepened* rather than *widened* (Hoogvelt 1997: 80), in the sense that its impact can be felt throughout the world, but in ways that have empowered some while increasing the vulnerability of others. It is not the case then that some parts of the world are effectively incorporated while others are insufficiently globalised; rather, it is that *the actual processes of globalisation that have occurred have been intrinsically uneven, unequal and unstable.*

The location of productive capital is similarly hierarchical. As Chapter 2 argues in some detail, even most transnational companies invest the bulk of their capital in their home country, and most of these same companies' foreign investment is in other 'advanced' capitalist countries. Moreover, recent tentative moves to new forms of flexible accumulation have reinforced this process, as companies rely on skilled labour or close proximity to suppliers and/or final markets (Kiely 1997: ch. 9; 1998b). In so far as Third World areas receive manufacturing foreign investment, much is for access to protected domestic markets. Alternatively, it may be labour intensive in nature and take the form of sub-contracting agreements with core companies, where most of the value added is at the design, distribution or marketing stage. Such practices are common in the clothing and textiles industries, but also in the assembly stage of high-tech production (Dicken 1992: chs. 8 and 10; Henderson 1989). As I argue in Chapter 2, these variations in the patterns of foreign investment are in part due to differences between locations and the ability of certain areas to meet particular locational requirements.

A key factor in promoting investment is the role of the state. Chapter 2 shows that the rise of the East Asian newly industrialising countries cannot be explained by neo-classical globalisation theories which suggest that South Korea, Taiwan and others industrialised on the basis of a comparative advantage in cheap labour. In fact, levels of foreign investment in these economies have been relatively low and the state has played a key role in forcing local capital to invest in manufacturing industry. Such capital has historically enjoyed high rates of protection from cheap imports as well as state subsidies (Kiely 1997: ch. 8). Similarly, the increase in East Asian foreign investment does not mean that the state ceases to play an economic role. British Conservative governments have argued that such investment confirmed their belief that in a global economy, nation-states could attract investment through a non-interventionist, 'free market' approach. In fact, cheap labour was only one of several attractions for foreign capital. More significant was the access such production gave to the European Union market[15] and the enormous subsidies (presumably not declared so that free market dogma could continue to flourish) effectively granted by British Conservative governments. Perhaps even more significant were the low *overall* rates of manufacturing investment in the 'deregulated' British economy as a whole (Harman 1996: 29).

The continued existence of core areas itself impacts on the movement of peoples across national borders. Unlike in the nineteenth century, people from the Third World have moved towards the West rather than vice versa.[16] In their attempts to continue to move to the global cities, such people are only too aware of where the 'cutting edge' of the global system is to be located. As Allen and Hamnett (1995b: 237) state: '[s]uch cities are not simply "on the map", their institutions are regarded by many to be primarily responsible for the economic map; that is, the networks of social relationships that lie behind a global economy characterized by geographical unevenness and inequality'. In Chapter 3, Phil Marfleet takes up this theme by challenging the standard distinction between economic migrant and political refugee, and showing how both political and economic factors combine to encourage movement away from the peripheries of the global system. He also shows how attempts to integrate peripheries into world capitalism through structural adjustment (discussed in detail in Chapter 1) have actually intensified their subordination, thus further encouraging migration. But as Marfleet shows, such processes of migration are themselves limited by the exclusionary practices, and hence continued vitality, of nation-states in the core countries. Indeed, the barriers to entry are greater now than they were 100 years ago (Hirst and Thompson 1996: 29–31).

Particular configurations of the global and local can also be found in the case of health. In recent years, a growing global awareness over health issues has developed, linked in particular to the spread of HIV/AIDS. In Chapter 4, Maureen Larkin examines some of the global dimensions of health and health policy in relation to the Third World. In comparing developments in health in nineteenth-century Britain with that of parts of the Third World today, she demonstrates the differences between the two experiences. In particular, the subordinate position of Third World countries within a global economy is emphasised, and the complex ways in which Western-driven policies for development and health interact and work to create new and uncertain prospects for health. These range from the impact of structural adjustment policies and how these have worked to exacerbate poverty and undermine basic prerequisites for health, to inadequately regulated work environments which expose populations to dangerous physical health hazards. Policies for health are also shown to be increasingly shaped by the interests and activities of a variety of global agencies such as transnational companies, multilateral agencies and consumer alliances. This is illustrated in relationship to policies on drugs, primary care and women's health.

Perhaps the main issue in which there has been a growing sense of global awareness in recent years is that of the environment. Destruction of rainforests has been linked to global warming and there has been substantial growth in international social movements concerned with the environment. This global awareness culminated in the 1992 United Nations Conference on the Environment and Development, held in Rio de Janeiro. At this Conference a new Convention on Biological Diversity was established, and Suzanne Biggs investigates the political issues around the issue of biodiversity in Chapter 5. She shows that any attempt to implement meaningful sustainable development must simultaneously deal with the inequalities between cores and peripheries in developing, controlling and utilising biotechnologies. In particular, Biggs demonstrates how transnational companies (TNCs) are globally restructuring the food and pharmaceutical sectors in ways that are likely to further undermine the economies of the poorest countries dependent on trade in agricultural commodities. TNCs are seeking to acquire recognition of Trade Related Intellectual Property Rights (TRIPs) which would give them ownership of any newly isolated genetic material used in their products, thus monopolising the future use of this biodiversity. These TRIPs are threatening internal markets and the use by peasant farmers of 'traditional' agricultural and pharmaceutical products in some countries, and they could threaten public access to plant genetic material held in the International Agricultural Research Centres. Unless subsequent conferences of the Biodiversity Convention find an equitable way to negotiate these critical environmental and development issues, an intensification of *uneven* rather than *sustainable* development is the likely outcome.

Such environmental inequalities can be seen also in more local cases. Desperate for foreign exchange, some poorer nations have used the absence of environmental regulations as a stimulant to attract foreign investment (Yearley 1995: 162–70). Too often, concerns with the environment lead to an avoidance of such inequalities and indeed can lead to an unfair portion of blame being placed on developing countries. This is remarkable when one considers that First World countries consume around 50 per cent of the world's energy resources compared to only one-sixth in the Third World, and that the former pumps out around 80 per cent of the world's greenhouse gases and 90 per cent of CFC gases that destroy the ozone layer (*New Internationalist* 1992: 18–19).

Culture

In some senses people across the world share a global culture. Again, new media technologies have played a crucial role in the promotion of television programmes, musical recordings and films for a global audience. However, for some writers (Petras 1993) the fact that these media industries are largely based in the West, and in particular in the United States, means that we need to ask 'whose culture?' Indeed, writers like Petras argue that the development of these global markets represents a form of cultural imperialism.

Concern over the 'Americanisation' of national cultures is not new. As well as debates in Nazi Germany, Stalinist Russia, post-war Britain and much of the Third World in the 1970s, there were earlier related fears of the vulgar tastes (or in more 'radical' analyses, the manipulability) of 'the masses' from the nineteenth century onwards (see Bennett 1982; Negus 1996: 203–6; Slater 1997: 64–74; Strinati 1992). More recently, the French film industry has been particularly concerned at the prospect of a 'free market' which would lead to the decimation of its national culture by 'cheap' American films (Morley and Robins 1995: 18).

The cultural imperialism thesis has rightly been subject to widespread criticism.[17] As Tomlinson (1991: ch. 4) points out, the thesis all too easily itself falls into a paternalism for the (assumed) passive dupes of the Third World; in other words, it patronises the Other. Related to this point, it falsely homogenises both the 'sending' and 'receiving' culture in any transaction. Not all American products present an account of that society which is sympathetic to the values and ideals of its dominant classes – one thinks here of US 'alternative rock', hip-hop and 'loser' television programmes such as *The Simpsons*, *Beavis and Butthead* and *Roseanne* (Kellner 1995: chs. 4 and 5; Straw 1997). Moreover, many American cultural products are made by non-US companies. For instance, of the 'Big Six' companies in the music industry (EMI, Polygram, Sony, Warner, BMG, MCA), only one is owned by a US parent company (Longhurst 1995: 30). Similarly, at the receiving end, do 'national cultures' break down simply because of the effects of foreign cultural products? This assertion has the effect of leaving the notion of national culture unproblematised and homogenised. This position ironically deprives such cultures of their own dynamic histories, which include interaction with many cultural traditions. The cultural imperialism thesis therefore replicates a crude economic–technological determinism by which cultural effects are simply 'read off' from economic processes.

Moreover, the culture industries do not work in such a crudely instrumental way. As Allen (1995: 117) argues, 'rather than eroding local differences, [global culture] actually works through them . . . [A]spects of local culture from wherever are taken up and *re-worked* within the global market-place, albeit one that is itself segmented by style, taste and, of course, money'. Thus globalisation does not promote complete global homogenisation – although this may be one tendency – as it can sell *difference* to segmented markets. Thus, in the core areas of the world today there is, for example, access to a wide variety of 'ethnic cuisine' and 'world musics'. These factors relate to the development of 'diaspora cultures', considered in some detail in Chapter 3. Such reworking may itself have contradictory effects, such as the sanitisation of cultural products in order to meet the demands of Western consumers eager to experience (what they believe to be) 'authenticity'.[18]

These questions are taken up in Chapter 6 by Vivian Schelling (and more briefly in a consideration of the marketing strategies of pharmaceutical companies in Chapter 4). In a discussion of the development of black culture in the city of Salvador, north-east Brazil, she shows how the development of a black identity was in part a struggle from below, but that the meaning of this identity was reworked as it was integrated into the global 'discourse of consumerism'. In further examples taken from Mexico, Schelling shows how in the case of artisans in Teotiltian del Valle, the affirmation of ethnic identity was used in part to gain greater control over the process of production.

The question of culture and globalisation is further examined by David Hesmondhalgh in Chapter 7. Taking a case study of the music industry, he reverses the usual direction of analysis, focusing more on the reception of Anglo-American-dominated music in other cultures, rather than the reception of Third World musics in the First. Whilst critical of crude interpretations of cultural imperialism, he argues that this does not mean that a level playing field has emerged. Instead, he argues that the logic of capitalism in general (as opposed to specific national capitalisms) encourages unequal access to the distribution and promotion of particular artists. He goes on to argue that the reception of 'Western' cultural goods will vary in meaning, and will depend on particular configurations of the local and the global. An interesting case study could be made, for example, between the reception of Western music before and after the fall of Stalinist rule in Eastern Europe (Pekacz 1994; Wicke 1992).

Hesmondhalgh's argument therefore parallels that of Shohat and Stam (1994), who criticise a tendency in cultural studies to champion a vacuous notion of resistance. This approach, most often associated with Fiske (1989), associates resistance on the part of consumers who create their own meanings out of various texts. As Shohat and Stam argue:

> Resistant readings, for their part, depend on a certain cultural or political preparation that 'primes' the spectator to read critically . . . [W]hile disempowered communities can decode dominant programming through a resistant perspective, they can do so only to the extent that their collective life and historical memory have provided an alternative framework for understanding.
>
> (Shohat and Stam 1994: 354)

The thesis that globalisation represents a simple extension of 'Westernisation' thus needs qualification. Nevertheless, the counter-examples chosen above still suggest that the West is in the driving seat. On the other hand, some writers (Robertson 1994) have identified the resurgence of Islamic activism as a global, but also anti-Western, force. The rise of 'global Islam' is thus regarded by Robertson as confirmation of his view, that globalisation universalises the particular and particularises the universal, referred to above.

In more official – and conservative – circles there has been a growing interest in the 'Islamic threat' and the perceived 'clash of civilisations' between the West and 'the Orient' (Huntington 1993). Although careful to distinguish between the two, Marfleet's consideration of globalisation theory and religion in Chapter 8 shows that both perspectives exaggerate the unity of 'Islam' and of religion more generally. Accounts of civilisational conflict and global religions falsely homogenise the great variety of movements in the world today. The rise of Islamic activisms and Liberation Theology can be explained partly by global *processes*, but such movements have their own *specificity*, based on particular relationships between the global and the local. Above all, the rise of such movements shows the continued vitality of *agency* and thereby undermines globalist accounts which read off such developments from the (structural–functionalist) logic of a supposed 'global level'.

Finally, the notion that globalisation simply represents the extension of Western domination has been undermined by the rise of East Asia. Some Western politicians have attempted to represent Japan in particular as the most dangerous Other, concerned, in the words of former French Prime Minister Edith Cresson, with 'world domination'. She has also argued that the Japanese 'stay up all night thinking about ways to screw the Americans and Europeans. They are our common enemy' (cited in Morley and Robins 1995: 147). Such blatantly racist comments tell us something about the way that many people have identified the modern with a specific geographical area, namely the West. The rise of new centres of capital accumulation thus once again shows that we must continue to think critically about space in the modern world. Moreover, it shows that globalisation has undermined many people's sense of national identity, despite the best efforts of parochial nationalists (Tomlinson 1996: 34).

Conclusion

This extended introduction has surveyed some of the main facets of the globalisation debate and shown how the rest of the book relates to these issues. I have been concerned to show the continued socio-political saliency of space, and in particular the inequalities that exist in the global order. In so doing, I have attempted to show that the world can still be divided into cores and peripheries, not in the sense that the former deliberately creates the latter, but in the sense that these divisions are an effect of the way that global social relations operate. Such cores and peripheries are therefore likely to change over time, but in such a way that uneven development will not be eliminated.

This brings us back to the question of the Third World. Worsley has argued that

> The Third World . . . is not a myth. The various meanings with which the term 'Third World' has been invested show family resemblances, even though they do not fully coincide. They have, that is, a common referent in

> the real world out there: the unequal, institutionalized distribution of wealth and illth on a world scale.
>
> (Worsley 1984: 339)

Thus, while non-alignment makes less sense in a post-Cold War world, and there is no economic unity (and probably never was) among an extremely diverse set of nations, the reality of global inequality means that, for Worsley, the Third World continued to exist. While such a definition lacks analytical rigour, and can easily fall back into a kind of timeless world system based on North–South inequalities, it at least has the merit of identifying such inequalities. The Third World therefore remains an important descriptive term, and it is in this sense that the contributors to this book continue to find it appropriate.

Thus, although the chapters that follow deal with a diverse set of areas of inquiry, they are all united around the following themes. First, that globalisation refers to processes that require concrete empirical investigation. Second, that these processes are real, but are in some sense limited and do not eliminate the importance of 'the local'. Third, that these processes have not eliminated, and are likely to intensify, uneven development; and that, therefore, globalisation, in so far as it exists, is differentially and unequally experienced in the world today.

Notes

1. The approach is therefore quite close to the Open University course 'The Shape of the World'. See Allen and Massey (1995), Allen and Hamnett (1995a), Sarre and Blunden (1995), Massey and Jess (1995) and Anderson *et al.* (1995).
2. The most notable exceptions are Sklair (1991), McMichael (1996) and Preston (1996). Interestingly, all of these writers have a background of teaching and research in development studies.
3. See for instance Ohmae (1991).
4. The concepts of postmodernity and postmodernism are particularly elusive. In the text I link them to the development of new communications technologies. In this sense 'post' means *after* modernity and is explicitly related to space. As well as the notion of a decentred hyper-space, where the real and imaginary are conflated, post-modernism also stresses the importance of difference, which includes attention to local space. However, there is a tendency in this approach to autonomise the local and ignore the global, a point I return to in the text below. Related to this point, postmodernity has been used to refer to a greater sense of *reflection* on the outcome of modernity, particularly its undesirable features (Bauman 1992: 102–3; Massey 1994: ch. 10). This is briefly referred to in the discussion of underdevelopment and postcolonial theories below.
5. In discussing Baudrillard, I am not suggesting that the globalisation writers mentioned above share his views. On the contrary, Giddens and especially Harvey have provided powerful critiques of extreme postmodern positions, such as those presented by Baudrillard. My purpose in discussing such an extreme position is to

show that local 'spaces' continue to exist in the face of globalisation, a position shared by both Giddens and Harvey.

6. A simple example taken from the British music industry illustrates the point. In an effort to get their acts to the top of the national (British Phonographic Industry) singles chart, record companies offer retailers certain incentives, such as 'buy one, get one free' or even better. As a result, the retailer can sell a single record cheaply in the first week and still make a profit. However, by the following week the record company has lost interest and markets another act. In the absence of special offers for the original single, the price increases and so it drops quickly out of the chart. Financial losses are recovered by sales in the lucrative album market. We therefore have a situation, according to record producer Pete Waterman, where '[i]t's no longer a music chart. It's a market chart. The media no longer believes it, the kids no longer believe it. It has no relevance' (*Guardian* 21 February 1997). This can be seen as a classic illustration of Baudrillard's position of the implosion of reality (in this case the national charts). For Baudrillard, it is not only the media (record companies), but also 'the masses' ('the kids') who reject the meanings attached to media images. However, the above example also shows the weakness of Baudrillard's approach, for there is clearly an *agency* (the record companies) and a relationship of power in this situation, even if this strategy may sometimes backfire (as not all albums sell). The implosion of meaning in this situation does not amount to 'mass resistance' to cultural domination, but in fact shows how record companies may openly cater to the implosion of meaning (Lash and Urry 1994: 289–90).
7. Most famously there were protests by Chinese students throughout the world in the wake of the suppression of the Tiananmen protest of 1989 – although even this was not without its contradictions as students in China often relied on sketchy information from the Western press (Penley and Ross 1993: viii–ix). The Internet is a useful source of up-to-date information on recent events such as the Zapatista uprising in Mexico and labour protests in South Korea. A useful summary of the history of the Internet, including its influence among a US counter-culture, can be found in Castells (1996: 345–64).
8. This is not true of all eighteenth- and nineteenth-century theories however. Many thinkers in the Enlightenment and post-Enlightenment tradition contained the seeds of knowledge that could be accommodated to the exercise of power, but also *resistance* to that power. This is perhaps most evident in the work of Marx.
9. Brief reference will also be made to the related world systems theory and dependency theory.
10. This term is taken from Wallerstein (1974).
11. Or, on the other hand, a blanket dismissal if the break with the West does not go far enough. Frank's writings in the 1970s, for example, were peppered with brief and all too sweeping assessments of the limitations of nationalist movements in the Third World, such as FRELIMO in Mozambique.
12. The above comments suggest that this approach *never* accurately represented the world. My point here is that the static model of underdevelopment can never account for changing relations of core and periphery.
13. This is not to deny that in *specific* cases, particular institutions may act in a conspiratorial way. A passing acquaintance with, for example, the Reagan years of

government in the United States shows that conspiracies exist which are partly designed to 'keep the Rest in its place'. However, no general theory of conspiracy can be assumed, and certainly global capitalism is too chaotic and disorganised to consciously implement underdevelopment. Hence my point that the periphery is an effect, rather than a deliberate cause, of the operation of the world economy.

14. More conservative theories are discussed in Chapter 1. The purpose of an extended critique of underdevelopment theory in this introduction should become clear in the text.
15. The issue of regionalism is addressed in Chapter 2.
16. But it should be remembered that most refugees actually move from one 'Third World' country to another (Zolberg *et al.* 1989).
17. Much of this cultural thesis parallels the economic claims made by underdevelopment theory.
18. On world music see Frith (1988). For an interesting reflection on these issues, and on the particularity of musical experience, see Toop (1997).

References

Ahmad, A. (1994) *In Theory: Nations, Classes, Literatures*, London: Verso.

Allen, J. (1995) 'Global worlds', in J. Allen and D. Massey (eds) *Geographical Worlds*, Oxford: Oxford University Press, pp.106–37.

Allen, J. and Hamnett, C. (eds) (1995a) *A Shrinking World? Global Unevenness and Inequality*, Oxford: Oxford University. Press.

Allen, J. and Hamnett, C. (1995b) 'Uneven worlds', in J. Allen and C. Hamnett (eds) *A Shrinking World? Global Unevenness and Inequality*, Oxford: Oxford University Press, pp. 234–52.

Allen, J. and Massey, D. (eds) (1995) *Geographical Worlds*, Oxford: Oxford University Press.

Amin, S., Arrighi, G., Frank, A. G. and Wallerstein, I. (1982) *Dynamics of Global Crisis*, New York: Monthly Review Press.

Anderson, J., Brook, C. and Cochrane, A. (eds) (1995) *A Global World? Re-ordering Political Space*, Oxford: Oxford University Press.

Baudrillard, J. (1981) *For a Critique of the Political Economy of the Sign*, St Louis: Telos.

—— (1983) *In the Shadow of the Silent Majorities*, New York: Semiotext.

—— (1988) *Selected Writings*, Cambridge: Polity.

—— (1991) 'The reality gulf', *Guardian* 11 January.

Bauman, Z. (1992) *Intimations of Postmodernity*, London: Routledge.

Bennett, T. (1982) 'Theories of the media, theories of society', in M. Gurevitch, T. Bennett, J. Curran and J. Woollacott (eds) *Culture, Society and the Media*, London: Methuen, pp.30–55.

Bernstein, H. (1979) 'Sociology of development versus sociology of underdevelopment', in D. Lehmann (ed.) *Development Theory: Four Critical Essays*, London: Frank Cass, pp.77–106.

Bernstein, H. and Nicholas, H. (1983) 'Pessimism of the intellect, pessimism of the will?', *Development and Change* 14: 609–24.

Bernstein, H., Hewitt, T. and Thomas, A. (1992) 'Capitalism and the expansion of Europe', in T. Allen and A. Thomas (eds) *Poverty and Development in the 1990s*, Oxford: Oxford University Press, pp. 168–84.

Bhaba, H. (1994) *The Location of Culture*, London: Routledge.

Booth, D. (1985) 'Marxism and development sociology: interpreting the impasse', *World Development* 13: 761–87.

Brenner, R. (1977) 'The origins of capitalist development: a critique of "neo-Smithian Marxism"', *New Left Review* 104: 25–92.

Castells, M. (1996) *The Rise of the Network Society*, Oxford: Blackwell.

Cowen, M. and Shenton, R. (1996) *Doctrines of Development*, London: Routledge.

De Vries, J. (1976) *The Economy of Europe in an Age of Crisis, 1600–1750*, Cambridge: Cambridge University Press.

Dicken, P. (1992) *Global Shift*, London: Paul Chapman.

Featherstone, M. (ed.) (1990a) *Global Culture*, London: Sage.

—— (1990b) 'Global culture: an introduction', in M. Featherstone (ed.) *Global Culture*, London: Sage, pp. 1–14.

Fiske, J. (1989) *Reading the Popular*, London: Unwin Hyman.

Frank, A. Gunder (1969) *Capitalism and Underdevelopment in Latin America*, New York: Monthly Review Press.

Frith, S. (ed.) (1988) *World Music, Politics and Social Change*, Manchester: Manchester University Press.

Frobel, F., Heinrichs, J. and Kreye, O. (1980) *The New International Division of Labour*, Cambridge: Cambridge University Press.

Giddens, A. (1990) *The Consequences of Modernity*, Cambridge: Polity.

—— (1991) *Modernity and Self-identity*, Cambridge: Polity.

Guardian, (1997) 21 February.

Gulalp, H. (1986) 'Debate on capitalism and development: the theories of Samir Amin and Bill Warren', *Capital and Class* 28: 135–59.

Hall, S. (1992) 'The West and the rest: discourse and power', in S. Hall and B. Gieben (eds) *Formations of Modernity*, Cambridge: Polity, pp. 276–320.

Harman, C. (1996) 'Globalisation: a critique of a new orthodoxy', *International Socialism* 73: 3–33.

Harris, N. (1986) *The End of the Third World*, Harmondsworth: Penguin.

Hart-Landsberg, M. (1979) 'Export-led industrialisation in the third world: manufacturing imperialism', *Monthly Review* 31: 181–93.

Harvey, D. (1989) *The Condition of Postmodernity*, Oxford: Blackwell.

Henderson, J. (1989) *The Globalisation of High Technology Production*, London: Routledge.

Hettne, B. (1995) 'Introduction: towards an international political economy of development', *European Journal of Development Research* 7(2): 223–32.

Hirst, P. and Thompson, G. (1996) *Globalisation in Question*, Cambridge: Polity.

Hoogvelt, A. (1997) *Globalisation and the Postcolonial World*, London: Macmillan.

Huntington, S. (1993) 'The clash of civilisations?', *Foreign Affairs* 72(3): 22–49.

Kellner, D. (1995) *Media Culture*, London: Routledge.

Kiely, R. (1994) 'Development theory and industrialisation: beyond the impasse', *Journal of Contemporary Asia* 24: 133–60.

—— (1995) *Sociology and Development: the Impasse and Beyond*, London: UCL Press.

—— (1997) *Industrialisation and Development: a Comparative Analysis*, London: UCL Press.

—— (1998a) 'Semi-periphery', in R. Jones (ed.) *Routledge Encyclopaedia of International Political Economy*, London: Routledge.

—— (1998b) 'Globalisation, post-Fordism and the contemporary context of development', *International Sociology* 13(1): 95–114.

King, A. (1995) 'The times and spaces of modernity (or who needs postmodernism?)', in M. Featherstone, S. Lash and R. Robertson (eds) *Global Modernities*, London: Sage, pp.108–23.

Lash, S. and Urry, J. (1994) *Economies of Signs and Space*, London: Sage.

Longhurst, B. (1995) *Popular Music and Society*, Cambridge: Polity.

Massey, D. (1992) 'A place called home?', *New Formations* 17: 3–15.

—— (1994) *Space, Place and Gender*, Cambridge: Polity.

Massey, D. and Jess, P. (eds) (1995) *A Place in the World? Places, Cultures and Globalization*, Oxford: Oxford University Press.

McMichael, P. (1996) *Development and Social Change*, London: Pine Forge.

Meyrowitz, J. (1986) *No Sense of Place: the Impact of Electronic Media on Social Behaviour*, New York: Oxford University Press.

Morley, D. (1996) 'EurAm, modernity, reason and alterity: or, postmodernism, the highest stage of cultural imperialism?', in D. Morley and K.-H. Chen (eds) *Stuart Hall: Critical Dialogues in Cultural Studies*, London: Routledge, pp.326–60.

Morley, D. and Robins, K. (1995) *Spaces of Identity*, London: Routledge.

Negus, K. (1996) *Popular Music in Theory*, Cambridge: Polity.

New Internationalist (1992) 'Green justice', no. 230.

—— (1996) 'Seduced by technology', no. 286.

Norris, C. (1992) *Uncritical Theory*, London: Lawrence & Wishart.

O'Brien, P. (1982) 'European economic development: the contribution of the periphery', *Economic History Review* 35(1): 1–18.

O'Brien, R. (1991) *Global Financial Integration: the End of Geography*, London: Pinter.

Ohmae, K. (1991) *The Borderless World*, London: Fontana.

Pekacz, J. (1994) 'Did rock smash the wall? The role of rock in political transition', *Popular Music* 13(1): 41–9.

Penley, C. and Ross, A. (eds) (1993) *Technoculture*, Minneapolis: University of Minnesota Press.

Petras, J. (1993) 'Cultural imperialism in the late 20th century', *Journal of Contemporary Asia* 23: 139–48.

Phillips, A. (1977) 'The concept of development', *Review of African Political Economy* 8: 7–20.

Preston, P. (1996) *Development Theory*, Oxford: Blackwell.

Radice, H. (1984) 'The national economy: a Keynesian myth?', *Capital and Class* 22: 111–40.

Robertson, R. (1990) 'Mapping the global condition: globalization as the central concept', in M. Featherstone (ed.) *Global Culture*, London: Sage, pp. 15–30.

—— (1991) 'Social theory, cultural relativity and the problem of globality', in A. King (ed.) *Culture, Globalization and the World System*, London: Macmillan, pp. 90–115

—— (1994) *Globalization: Social Theory and Global Culture*, London: Sage.

Rodney, W. (1974) *How Europe Underdeveloped Africa*, London: Bogle L'Ouverture.

Said, E. (1978) *Orientalism*, Harmondsworth: Penguin.

Sarre, P. and Blunden, J. (eds) (1995) *An Overcrowded World? Population, Resources and the Environment*, Oxford: Oxford University Press.

Shohat, E. and Stam, R. (1994) *Unthinking Eurocentrism: Multiculturalism and the Media*, London: Routledge.

Sklair, L. (1991) *Sociology of the Global System*, Brighton: Harvester-Wheatsheaf.

Slater, D. (1997) *Consumer Culture and Modernity*, Cambridge: Polity.

Stallabrass, J. (1995) 'Empowering technology: the exploration of cyberspace', *New Left Review* 211: 3–32.

Stavrianos, L. (1981) *Global Shift*, New York: Morrow.

Straw, W. (1997) 'Communities and scenes in popular music', in K. Gelder and S. Thornton (eds) *The Subcultures Reader*, London: Routledge, pp. 494–505.

Strinati, D. (1992) 'The taste of America: Americanization and popular culture in Britain', in D. Strinati and S. Wagg (eds) *Come on Down? Popular Media Culture in Post-war Britain*, London: Routledge, pp.46–81.

Taylor, P. (1992) *War and the Media: Propaganda and Persuasion in the Gulf War*, Manchester: Manchester University Press.

Thrift, N. (1995) 'A hyperactive world', in R. Johnston, P. Taylor and M. Watts (eds) *Geographies of Global Change*, Oxford: Blackwell, pp.18–35.

Tomlinson, J. (1991) *Cultural Imperialism*, London: Pinter.

—— (1996) 'Cultural globalisation: placing and displacing the West', *European Journal of Development Research* 8(2): 22–35.

Toop, D. (1997) 'The meaning of world music', *The Wire* 155: 74.

Vincent, R. (1997) 'The future of the debate: setting an agenda for a New World Information and Communication Order, ten proposals', in P. Golding and P. Harris (eds) *Beyond Cultural Imperialism*, London: Sage, pp. 175–207.

Wade, R. (1996) 'Globalization and its limits: reports of the death of the nation-state are greatly exaggerated', in S. Berger and R. Dore (eds) *National Diversity and Global Capitalism*, Cornell: Cornell University Press, pp.60–88.

Wallerstein, I. (1974) *The Modern World System*, New York: Academic Press.

Wallerstein, I. (ed.) (1983) *Labour in the World Social Structure*, Beverly Hills: Sage.

Walsh, J. (ed.) (1995) *The Gulf War Did Not Happen*, Aldershot: Arena.

Waters, M. (1995) *Globalization*, London: Routledge.

Wicke, P. (1992) 'The role of rock music in the political disintegration of East Germany', in J. Lull (ed.) *Popular Music and Communication*, London: Sage.

Worsley, P. (1984) *The Three Worlds*, London: Weidenfeld & Nicolson.

Yearley, S. (1995) 'Dirty connections: transnational pollution', in J. Allen and C. Hamnett (eds) *A Shrinking World? Global Unevenness and Inequality*, Oxford: Oxford University Press, pp. 143–82.

Zolberg, A., Suhrke, A. and Sergio, A. (1989) *Escape from Violence*, Oxford: Oxford University Press.

chapter 1

The crisis of global development

Ray Kiely

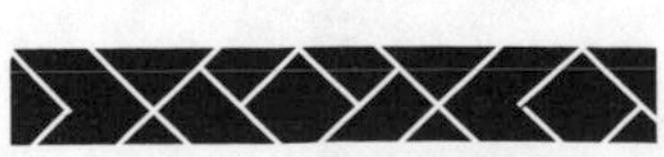

Whilst the 1980s and 1990s can be considered a time of crisis, the period from the 1950s to the 1970s was one of great optimism concerning the prospects for development in the Third World. The 1960s was declared the United Nations' First Decade of Development, with the declared aim to be 'the attainment in each less developed country of a substantial increase in the rate of growth, with each country setting its own target, taking as the objective a *minimum* annual rate of growth of aggregate national income of 5 per cent at the end of the decade' (*UN Yearbook* 1961, cited in Leys 1996: 109). This growth rate target was increased to 6 per cent per annum in the Second Development Decade (the 1970s), and ambitious industrialisation targets were also set. These growth rate targets were far higher than those achieved by most of the earlier industrialised countries. Perhaps even more important, development practitioners generally believed that such growth rates would lead to the alleviation of poverty in the developing countries.

The 'lost decade' of the 1980s saw the apparent death of such optimism. By the beginning of the 1990s, most people in sub-Saharan Africa were poorer than they had been thirty years before. Of the population of about 500 million, nearly 300 million are living in absolute poverty (Leys 1994: 34). In the developing countries as a whole, nearly 800 million people do not get enough food, and about 500 million are chronically malnourished. Almost one-third of the population of developing countries – about 1.3 billion – live below the poverty line. The infant mortality rate, at around 350 per 100,000 live births, is about nine times higher than that in the 'advanced' industrial countries (UNDP 1995: 16).

Clearly, then, 'development'[1] has in some sense failed. Growth rates for many parts of the Third World – parts of East Asia and Latin America were notable exceptions – rarely reached the projected figures outlined at the start of the development decades, and in many cases fell in per capita terms in the 1980s. Moreover, the growth that did occur in the 1960s and 1970s often failed to 'trickle down' and so did not appear to relieve the problem of poverty. These problems were intensified in the 1980s as growth rates and living standards for many fell.

This chapter examines the crisis of development in the 1990s. The principal concern is therefore with discourses of development and the Third World, rather than with globalisation – although it will become clear that after 1945 the development discourse was a global one, and debates around development were increasingly simultaneously debates about the character of global processes. In focusing on development the chapter stresses its uneven, unequal and (in part) contingent and contradictory nature, issues taken up in later chapters which are more explicitly concerned with globalisation processes.

I start by outlining the context and content of development, and how this has shifted over time. This is followed by a critical analysis of the dominant development discourse of the 1980s and 1990s, neo-liberalism. I then attempt to deconstruct the idea of development, and criticise its tendency to reduce concrete agents of development to passive objects, awaiting the benefits of development. At the same time, however, I show that a simplistic anti-development position is itself problematic. Finally, I conclude by suggesting that attempts to impose development in a top-down fashion has led to resistance in the Third World (and elsewhere), an issue taken up in later chapters.

The context and content of development

Although the idea can be traced back to at least the nineteenth century, it was in the post-war period that the idea of development was made explicit. In his inaugural address as president to the people of the United States in 1949, Truman argued that there was a need for the countries of the modern world to solve the problems of the underdeveloped areas.

> More than half the people of the world are living in conditions approaching misery. Their food is inadequate, they are victims of disease. Their economic life is primitive and stagnant. Their poverty is a handicap and a threat both to them and to more prosperous areas. For the first time in history humanity possesses the knowledge and the skill to relieve the suffering of these people . . . I believe that we should make available to peace-loving peoples the benefits of our store of technical knowledge in order to help them realize their aspirations for a better life . . . What we envisage is a program of development based on the concepts of democratic fair dealing . . . Greater production is the key to prosperity and peace. And the key to greater production is a wider and more vigorous application of modern scientific and technical knowledge.
>
> (cited in Escobar 1995a: 3)

Truman's famous address is a very clear statement of the basic thinking behind the idea of development. The Third World (as it came to be known) was regarded as backward and primitive, but these problems could be overcome by following a similar path of development to that of the Western (civilised) world. Indeed, this path would be achieved more easily because the West could share the benefits of material prosperity and scientific knowledge with the 'backward' areas, and so hasten the transition to modernity. Above all, this could be achieved through an increase in production in underdeveloped areas, and this in turn could occur through the introduction of rational scientific methods.

This section examines the *context* within which post-war development thinking emerged, and the concrete *content* of development strategies.

The context of development

The idea of development arose in a very specific post-war context. The United States had emerged as by far the most powerful economy in the world, and so it was bound to have a strong influence on the structure of the global economy. At the same time, there was an alternative social and political system to that of the

West – communism in the Soviet Union. Equally important was the fact that it was clear by the end of the war that the days of the old European and Japanese empires were numbered, as people in the colonial world demanded independence. These three inter-related factors – US dominance, the Soviet alternative and the beginning of the end of empires – all influenced the way in which development was perceived.

At the end of the Second World War, the United States controlled around 70 per cent of the world's gold and foreign exchange reserves, and 40 per cent of industrial output (Brett 1985: 63). By this time the United States was largely converted to the principle of free trade (although certain interests such as the farming lobby successfully resisted its full implementation both nationally and internationally), but such a commitment was bound to be resisted by weaker economies who feared being out-competed by the powerful US economy. At the same time, the United States was also committed to an anti-communist foreign policy, and it was primarily for this reason that it was prepared to compromise with other countries over the shape and future direction of the international economy. The institutions which were to influence the nature of the post-war international economy were established at Bretton Woods in the United States in 1944, and they largely reflected the US interest in free trade combined with a willingness to compromise with its allies. As Brett (1985: 63) argues, the Bretton Woods agreement was based on compromises by the United States 'in which it used its economic strength to provide short-term inducements to the weaker countries to co-operate, in exchange for a willingness to build a long-term commitment to liberalisation into the structure of the institutions themselves'. There were only eighteen developing countries present at the Conference out of a total of forty-four countries, and so not surprisingly the interests of these nations were largely marginalised.

At Bretton Woods the International Monetary Fund (IMF), the World Bank and the General Agreement on Tariffs and Trade (GATT) were established. The principal role of the IMF was to regulate the economies of countries facing balance of payments deficits. After Bretton Woods, exchange rates were fixed by tying currencies to the value of the dollar, whose value was in turn fixed to the value of gold. It was recognised, however, that some countries facing balance of payments deficits would have to devalue in order to return to a situation of equilibrium. The IMF was to play a crucial role in regulating this process through the provision of short-term credit to countries in this situation. Some representatives (such as the British economist John Maynard Keynes) at Bretton Woods recommended that large amounts of credit be supplied to the IMF, largely from the 'advanced' industrial countries. The provision of such credit had the potential to transfer resources from the richer to the poorer countries, a position not dissimilar (although considerably milder) to the Third World demands for a new international economic order in the 1970s (discussed below). The United States, however, successfully resisted this proposal, and the IMF was left with limited financial resources. As a result, the conditions applied to IMF loans were often very stringent, and this became a major (possibly *the*) development issue in the 1980s. The resources were paid by the member states of the Fund, and voting power reflected the size of the contribution made by member states – and was therefore heavily weighted towards the industrial nations.

Keynes similarly wanted the World Bank to have sufficient amounts of credit for nations to borrow from it at cheap and affordable rates, but the US view again won the day. The Bank received little money from governments and so it had to raise most of its funds from private financial institutions. The result was that the Bank, set up to promote development in the 'underdeveloped areas', had to charge commercial rates of interest to borrowers and promote conservative financial policies. Government contributions were significant in so far as the size of each contribution determined voting strength within the institution. Some later concessions were made to the Keynesian view, such as the creation of the International Development Association, a soft loan affiliate of the Bank, in 1960. However, this institution similarly lacked the funds which could have made softer loans more generally available.

Finally, there was the General Agreement on Tariffs and Trade, which was less a formal institution than a debating forum for international trade negotiations. A more formal International Trade Organisation was rejected by the US Congress, largely in response to the interests of domestic US farmers, who opposed free trade in agriculture. The GATT was committed to the gradual implementation of free trade, through, for instance, the promotion of a general trend in tariff rates, but developing countries were allowed some concessions – for example in terms of effective rates of protection for domestic industry. These concessions, which increased in the Kennedy Round of tariff negotiations, 1964–7, encouraged increasing numbers of previously sceptical developing countries to become signatories to the GATT.

On the whole, then, the post-war settlement was favourable to the United States, but it was also based on unavoidable compromise. The United States, fearing the spread of communism, was prepared to give sufficient space (hence some protectionism) and indeed actively encouraged (through Marshall Aid) the 'advanced' capitalist countries to rebuild their economies. The developing countries were in a minority at Bretton Woods (most were still colonies in 1944), and so their voice was largely unheard. However, it was clear by this time that the days of empire were numbered as nationalist movements demanded independence throughout the colonial world. The United States for its part, fearing a spread of communism if Third World independence movements were ignored, was largely sympathetic to these calls. The nationalist movements which won independence for new states throughout the world were for their part largely sympathetic to the idea of 'development', although they had their own ideas about how this might be brought about.

The content of development

As the quotation from Truman above makes clear, the key to the development of the underdeveloped areas was economic growth. Development was regarded as a technocratic process in which the state would play a leading role in planning output and investing in dynamic sectors such as new industries. In this way, the underdeveloped areas would gradually catch up with the advanced industrial world. Development – or modernisation – was therefore defined as

> the process of change toward those types of social, economic and political systems that have developed in Western Europe and North America from the seventeenth century to the nineteenth and have then spread to other European countries and in the nineteenth and twentieth centuries to the South American, Asian and African continents.
>
> (Eisenstadt 1966: 1)

Although there were some important differences of emphasis between different theorists and practitioners of development, all were united around the strategy of modernisation through rapid economic growth. Industrialisation was seen as a key strategy because new rounds of technological investment increase productivity and thereby expand output. As Moore argued:

> What is involved in modernization is a 'total' transformation of a traditional or pre-modern society into the types of technology and associated social organization that characterize the 'advanced', economically prosperous, and relatively politically stable nations of the western world.
>
> (Moore 1963: 89)

The leaders of new states in the Third World were critical of the way in which the existing international order tended to marginalise their interests, and responded by forming pressure groups that attempted to change Northern priorities and the balance of power in the world system. In 1961 the Non-Aligned Movement was formed in order to promote an independent path between the interests of the communist and capitalist world. This was followed in 1964 by the formation of the United Nations Conference on Trade and Development (UNCTAD), which attempted to win some reforms in the international economy. These organisations had a different emphasis from Western-based theories of modernisation, but both approaches shared the assumption that the most effective strategy for development was rapid economic growth. Nationalist leaders such as Nehru (India), Nkrumah (Ghana) and Sukarno (Indonesia) were particularly keen to develop the potential of their people, which had been held back by colonialism, while at the same time taking advantage of the opportunities that Western aid and technology had to offer.

The major development strategy used in the Third World from the 1950s to 1970s was import-substitution industrialisation. This policy involved the development of a domestic industrial sector, initially producing for the domestic market, but with the long-term aim of breaking into lucrative export markets. In order to develop these industries, the state would have to protect new producers from competition from cheaper foreign imports, through high tariffs or import controls. Although initial investments were to be concentrated in consumer goods industries, state planners envisaged a time when new states would diversify into the intermediate and capital goods sectors such as industrial machinery.

Post-war development was therefore based on state planning, with the explicit purpose of raising productivity and output. Such planning applied not only to the industrial sector, but also to strategies designed to increase output in agriculture such as the Green Revolution. This strategy involved the introduction of a technological package such as modern high-yielding varieties, chemical fertilisers and irrigation schemes with the express purpose of increasing agrarian output. This strategy was actively implemented by many Third World states (such as Mexico, the Philippines and India) from the 1950s and 1960s onward, and enjoyed the backing of pro-development institutions from the United States such as the Rockefeller Foundation.

From the late 1940s to the late 1960s, then, development was largely seen as a process whereby 'the traditional Rest' caught up with 'the modern West'. This was to be achieved primarily through economic growth (although the adoption of modern Western values was also considered important – see McClelland 1961), which would raise productivity. The resultant increase in output would relieve the 'handicap of poverty' and thereby enable the peoples of the Third World to 'realize their aspirations for a better life'.

By the late 1960s, however, such a view was becoming increasingly difficult to sustain. International Labour Office (ILO) figures for 1972 estimated that 39 per cent of the population of the Third World was 'destitute', and 67 per cent was described as 'seriously poor' (cited in Kitching 1982: 70). Such figures showed the inadequacy of official measures of development, such as Gross National Product. GNP measures the total output of a country for a given year, and this is usually divided by the size of the population in order to arrive at a per capita GNP figure. The problem, however, is that such a figure tells us nothing about the way that such income is distributed within a country, or whether everyone has access to elementary needs such as food, education, health-care and housing.

Recognition of these problems helped to pave the way for a revised development strategy in the 1970s, based on the ideas of 'redistribution with growth' and 'basic needs'. The ILO first put forward the idea of redistribution with growth in 1972, while the World Bank adopted its basic needs strategy the following year. These strategies argued that growth should remain a priority of development strategies, but that this should be combined with increased attention to those (basically the poor) who had previously been marginalised. As the World Bank argued:

> Past strategies in most developing countries have tended to emphasize economic growth without specifically considering the manner in which the benefits of growth are to be redistributed . . . Although in the long run, economic development for the growing rural population will depend on expansion of the modern sector and on non-agricultural pursuits, too strong an emphasis on the modern sector is apt to neglect the growth

> potential of the rural areas. Failure to recognize this has been a major reason why rural growth has been slow and rural poverty has been increasing.
>
> (World Bank 1975: 16)

As this quotation makes clear, particular attention was paid to the small farmer and the need for rural development. The World Bank's president at the time, Robert McNamara, was principally concerned with increasing the productivity of the small farmer, so that 'they duplicate the conditions which have led to very rapid agricultural growth in a few experimental areas . . . so as to stimulate agricultural growth and combat rural poverty on a broad scale' (McNamara 1975: 90). Redistribution with growth also paid attention to developing the productive potential of the masses of people in the towns who were not in formal employment. The ILO in particular placed much emphasis on developing appropriate labour-intensive technologies, which would both enhance labour productivity and ensure some 'trickle down' of wealth as more people were employed (see Kitching 1982: 70–84).

These strategies were hindered to some extent by a lack of clarity over aims and methods, but perhaps more importantly by a failure to confront the realities of power within different parts of the world (Leys 1975). Put simply, many states in the Third World had little interest in alleviating the suffering of the 'seriously poor'. Moreover, those making investment decisions in the Third World (be they transnational companies or local capitalist enterprises) were less interested in securing full employment than in winning high rates of return on their investment, which in many cases involved the use of labour-displacing technology. In any case, by the early 1980s the tide had turned against such strategies, and development entered a new phase.

Contemporary development: neo-liberalism and the role of market forces

The neo-liberal 'counter-revolution' in development studies (Toye 1987) arose in the context of the debt crisis of the early 1980s. Briefly, the Bretton Woods system of fixed exchange rates was effectively abandoned between 1971 and 1973, when the United States devalued the dollar against the price of gold and gradually allowed for the introduction of a system of floating exchange rates in the world economy. Although there was still some attempt to regulate the exchange rates of particular nations (for instance, through the IMF), there was no successful replacement for the Bretton Woods system (for more details on the collapse of fixed exchange rates, see Brett 1985: 111–25).

This partial collapse of managing the world economic system coincided with a massive rise in the price of oil (1973–4), which led to windfall profits for oil-producing countries. These profits were largely deposited in Western banks, who then loaned them to selected Third World countries (mainly in Latin America) at low rates of interest. An effective 'free-for-all' developed by which the banks lent enormous amounts of money to a handful of countries – by 1982, the nine largest US banks had lent over twice their combined capital base to non-oil-producing Third World countries (Gibson and Tsakalatos 1992: 51).

In 1982, faced with increasing interest payments on its debts, Mexico declared itself unable to meet its debt obligations. Other countries followed and so the banks responded by attempting to stop new lending. A debt crisis had emerged in which international financial institutions faced the prospect of not receiving interest payments due to them and therefore they faced bankruptcy. This was avoided by the actions of national governments, which protected the most powerful banks from bankruptcy (although many smaller banks, particularly in the United States, went under), and by the role of the IMF in regulating the debt crisis. This context was reinforced by the rise of governments committed to neo-liberal reforms (both nationally and internationally), such as Margaret Thatcher's Conservative government in Britain, and Ronald Reagan's Republican regime in the United States. Given this context, it was not surprising that institutions such as the IMF and the World Bank began to advocate policies that more or less reflected the neo-liberal paradigm.[2]

The main contention of neo-liberal theory is that 'orthodox' state-led development, and in particular import-substitution industrialisation (ISI), is inefficient. State intervention in the economy was deemed to be inefficient for three principal reasons. First, state protection of domestic producers had the effect of reducing competition in the economy. The result was that there was no incentive for companies to be efficient, and so the state was in effect protecting inefficient producers. This in turn meant that the goods that were produced were of a high cost, and low quality, and these costs were passed on to domestic consumers within particular nation-states. Second, state intervention encourages unproductive 'rent-seeking'. This activity can be broadly defined as unproductive income-earning economic activity derived from state regulations (Kreuger 1974). For example, state controls over imports through licensing leads to lobbying pressure by companies to secure access to one of the scarce licences. Companies are thereby encouraged to spend a good deal of their time on lobbying for a licence (which may involve bribing government officials) rather than on productive, wealth-creating activity. Third, state regulations have the effect of automatically discriminating against some sectors at the expense of others. For instance, a state-protected and over-valued exchange rate has the effect of making potential exports more expensive in the world market, and so acts as a disincentive against would-be exporters operating within a particular nation-state (World Bank 1984: 35).

Neo-liberals therefore argue that ISI was not a sustainable strategy for Third World developers because it encouraged the growth of inefficient, expensive activity (through state protection of domestic producers); discouraged export-earning activity (through states artificially maintaining high exchange rates); and discouraged traditional wealth-creating (and also export-earning) activity such as farming (through state discrimination in favour of industry and against agriculture). It is not surprising, therefore, neo-liberals contend, that many nations found themselves in a situation by the early 1980s in which they were importing far more than they were exporting.

The neo-liberal remedy for these problems was the encouragement of the private sector and the liberalisation of Third World economies. Three key policy proposals were recommended: currency devaluation, rolling back the state and the liberalisation of international trade. Devaluation of the national currency is said to encourage exports by making them cheaper on the world market; and at the same time discourages over-reliance on imports by making them more expensive (World Bank 1984: 35). Rolling back the state entails a reduction in state economic activity such as planning, and the consequent promotion of 'unhindered market forces' as the best route to development. The encouragement of market forces encourages the most efficient means of allocating resources, and enables potential investors (and consumers) to respond to the correct price signals, thereby stimulating effective economic activity. Third, the liberalisation of international trade entails the abolition of import controls, and the reduction in tariff rates, which will force firms to be efficient as they face competition from foreign producers. On the whole, then, neo-liberalism maintains that state-led development planning has failed, and the remedy involves allowing market forces to operate unhindered by inefficient regulations. Each country in the world economy can then concentrate on producing goods cheaply and efficiently; each country thereby specialises in producing those goods in which they have a *comparative advantage*. Country A exchanges the products of its comparative advantage with the products of Country B's comparative advantage, and so each country benefits from the production and consumption of more goods. The World Bank has claimed, albeit with some qualifications, that a commitment to market-friendly policies was the main reason for the development successes in East and South-East Asia in the 1980s (World Bank 1993).

Neo-liberal theory has had a strong influence on the structural adjustment and stabilisation programmes introduced by many governments (with IMF and World Bank advice) since the early 1980s. Stabilisation programmes promoted by the IMF are designed to alleviate short-term balance-of-payments deficits and typically involve currency devaluation and public-spending cuts. Structural adjustment programmes, designed to promote long-term economic development, involve policies such as liberalisation and privatisation, as well as attempts to introduce competition and user costs into the state sector. The impact of adjustment has varied across countries (and economic sectors within countries) and it would be mistaken to assume that adjustment policies are simply the product of messianic World Bank neo-liberals imposing their will on weak Third World governments. In practice, adjustment policies have involved long and complex negotiations between the World Bank and Third World states, and have varied in their intensity according to the impact of recession within a particular country, and domestic political factors (Mosley *et al.* 1991).

Nevertheless, a certain 'family resemblance' can be found between different adjustment policies in different countries (Toye 1994: 29). Common policies include the removal of import quotas, an increase in export incentives, a revision in pricing policy for agricultural goods, and reform of the national budget and system of taxation. The impact of adjustment policies has on the whole not been as successful as the World Bank envisaged, and this can in part be attributed to their weak theoretical basis (Mosley 1994). Neo-liberalism is characterised by an excessive optimism concerning the role of market forces in promoting development. It assumes that nation-states can relatively easily break into export markets on the basis of their 'comparative advantage', but the reality is somewhat different. Producers in the Third World in particular face non-tariff protectionist barriers in First World markets, and even in a situation of free trade, competition often remains unequal. First World producers monopolise the most advanced technology, research and development, marketing practices and so on. This occurs in a wide variety of products, which in turn enables advanced country producers to develop substitutes for some products (such as corn fructose syrup instead of sugar cane). The result is that Third World producers often face strong constraints when attempting to break into new markets. The number of established products originating from the Third World is insufficient to enable developing countries to compete on an equal basis with the First World, with the consequence that most international capital flows are between First World countries (see Chapter 2). This does not mean of course that it is *impossible* that new export markets can be developed by particular countries, but neither does it mean that such an occurrence is *inevitable*.

Unequal competition also exists *within* as well as *between* countries. Late developers attempting to develop new production outlets in a free market environment will face the prospect of competition from more established producers from overseas. It is for this reason that many countries have advocated state protection of new industries – a practice that was (and remains) common in the newly industrialising countries in East Asia (Wade 1990). Adjustment policies such as the blanket removal of import controls and state subsidies, and tariff reduction, can therefore be regarded as inefficient, even from the narrow viewpoint of promoting economic growth (Mosley 1994).

Market forces may not only be inefficient, they may also promote injustice. Markets are hierarchical institutions in which some people enter into transactions on a far from equal basis with others. As Mackintosh (1990: 50) argues, the 'profits of a few, and growth for some, thrive in conditions of uncertainty, inequality and vulnerability of those who sell their labour power and of most consumers'. Thus, the owners of private property can hire workers who have only their labour power to sell, and then dispense with them when they are no longer required. This leaves such workers in a position of vulnerability, reliant on employers for access to a wage in order to buy basic necessities. State intervention constituted, however imperfectly, a partial attempt to relieve the situation of sections of the poor in the Third World – through for instance the provision of subsidies on basic foods. In some cases, the removal of such subsidies has led to riots, protesting against the actions of both Third World governments and international institutions (Walton and Seddon 1994).

The neo-liberal era of development has not been a success, and in some respects has actually made things worse. The figures cited at the beginning of this chapter show that the promises of development, in whatever form, have not been fulfilled. Could it be that development is not the solution but is in fact the *problem*?

Development 'deconstructed'

The history of the idea of modern development can be traced back to the emergence of industrial capitalism in eighteenth-century Europe (Cowen and Shenton 1996). Development was 'imagined' as a way to stabilise the chaos caused by the industrial revolution in Europe, 'to create order out of the social disorder of rapid urbanization, poverty and unemployment' (*ibid.*: 29). Development therefore has a far longer history than its supposed invention by President Truman in the 1940s. Nevertheless, the idea of development underwent a 'sea change' in the post-war era as newly independent peoples in the Third World came to see themselves in terms of the idea of development (Escobar 1995b: 213–14).

This section attempts to show that the optimism of post-war development thinking was misplaced. There were motives other than altruism which influenced the policy of the United States and its allies towards the Third World, and the ideology of development – at least as conceived by Truman and like-minded thinkers – was full of contradictions. From the point of view of the 'backward areas', then, both the context and the content of development were problematic.

The context of development

The last section made clear that the post-war agreement at Bretton Woods largely marginalised the so-called developing world, and instead a world order was constructed that largely reflected the interests of the United States and its allies. This meant that the Third World faced an unfavourable context for development to occur. Although parts of the developing world received direct foreign investment and aid, this was never substantial enough to redress the inequalities that existed in the global order. Most direct foreign investment flowed between the advanced capitalist countries (see Chapter 2), and aid was often used as a political or economic weapon by dispensers. The United States, for example, was notorious for dispensing aid to Cold War allies rather than to those most in need, and at times was even prepared to use food aid as a weapon to influence government policy in the Third World.[3] The dispensing of aid has been tied also to recipients buying goods from the donor country of origin (Bennett 1987). The developing world also faced some problems in the terms on which they traded with the First World. Third World primary goods producers often found that the prices paid for their products were volatile, and subject to sharp falls, and that there was often limited demand for their products.[4]

Organisations such as UNCTAD pressurised the economically powerful nations to reconstruct the international economic order in ways which were more conducive to the interests of developing nations. In the 1970s, the United Nations' General Assembly (where Third World nations had a majority) voted for a new international economic order, which would provide for more aid and investment (with fewer strings attached), codes of conduct for transnational company investment and measures to alleviate unequal terms of trade such as price compensation schemes or fixed prices for primary products (Anell and Nygren 1980). These proposals were not without their problems and inconsistencies (Toye 1983), but more important for our purposes was the fact that they remained largely ignored by the dominant powers. This indifference increased in the late 1970s and early 1980s with the neo-liberal counter-revolution.

The context of post-war development was thus one in which: (a) at best, the power of political and economic forces undermined any good intentions that may have existed in the 'development community'; (b) at worst, it was subordinated to the interests of the Cold War, global capital and so on. This has been made all the more stark in recent years with the neo-liberal revival. Although Cold War interests have not been so great, the triumph of neo-liberalism can be seen as a reflection of dominant interests in the global economy. Leys makes this clear:

> What made possible the triumph of neo-liberalism in mainstream development thinking was material, not ideal: the radical transformation in both the structure and the management of the world economy that had begun in the 1960s, and which finally seemed to offer the possibility of creating for the first time in history a truly unified global capitalist economy – and one regulated, if at all, only by institutions reflecting the interests of transnational capital. Neo-liberalism articulated the goals and beliefs of the dominant forces that stood to benefit from this process, and pushed it forward.
>
> (Leys 1996: 19)

The content of development

The problems run deeper than simply an unfavourable context, however, and relate to the very idea of development. One writer (Esteva 1987: 135) has even gone so far as to assert that 'You must be either very dumb or very rich if you fail to notice that development stinks.' Some writers (Marglin 1991; Sachs 1992) argue that the idea of development is simply a Eurocentric one, a form of cultural imperialism. The Third World is regarded as a cultural construction by the West, in some sense lacking what the Western developed world has. The former therefore needs to become more 'like us', and in the post-colonial era, development is the means by which this process can occur. Development is therefore designed to serve entrenched interests and/or to reinforce Western dominance in the global order, a form of 'arrogant interventionism' (Sachs 1992: 2).

Can this argument, that development is simply a form of cultural imperialism, be sustained? There is no easy answer to this question, but some weaknesses with the 'anti-development' position should be noted. First, a simple dismissal of development as Eurocentrism is too simplistic. The argument homogenises both the West and Third World, and reduces the latter to passive recipients of the former's ideas. It then becomes difficult to imagine how the people of the Third World could behave in any way other than being simple puppets of the West. An adequate analysis of development would regard it not only as a European creation, but also as a reflection of 'the responses, reactions and resistance of the people who are its object' (Crush 1995a: 8). The anti-development thesis similarly reduces development to an idea without history, impervious to change. In fact, development discourse – its language, strategy and practice – has changed over time, both in response to previous strategies and to shifts in power relations in the real world.

Second, the problems of development notwithstanding, the anti-development thesis focuses solely on the 'dark side' of development, and fails to take account of the fact that for all its faults, development can be empowering. The World Bank (1994: 393) has recently claimed that '[d]uring the past fifteen years, the share of households with access to clean water has increased by half, and power production and telephone lines per capita have doubled. Such increases do much to raise productivity and improve living standards.' Of course, this still leaves many without these services, but anti-development theory seems ill-prepared to discuss these questions, especially as it focuses solely on the ill-effects rather than the possibilities of technology (see Ullrich 1992). An alternative, more realistic perspective would 'recognize that development is about costs and benefits, and [be] . . . more interested in the balance of these items than in the possibility of a painless development or non-development' (Corbridge 1995a: 10).

Third, anti-development positions tend to be rather vague about alternatives and romanticise local cultures (Nanda 1991; Kiely 1995b). The idea that development simply undermines authentic cultures (see Shiva 1989, 1991) can itself be considered a form of cultural imperialism, based on a long tradition of romanticising 'the Other' (Said 1978). In fact, the idea of a 'crisis in development' is as old as the idea of development itself. This can be seen for instance in the populist alternatives to the chaos of European industrialisations (Kitching 1982: chs. 2 and 3; Watts 1995). Similarly, the argument that needs and poverty are *solely* social and cultural constructions (see Rahnema 1991) 'mistake(s) the word for the World . . . [and] can create the conditions wherein a positive local politics of empowerment slides fitfully into an amoral politics of indifference' (Corbridge 1994: 97).

Furthermore, in attempting to preserve that which is willfully destroyed by development (see Marglin 1991), anti-development fails to take us beyond the dualisms of modernity and tradition, and dominant and dominated (Pieterse and Parekh 1995: 8–9; Cowen and Shenton 1996: 467–70). This is an important point with serious political implications, for as Manzo (1995: 238) states: 'Efforts in the post-colonial world to reinvent a pre-colonial Eden that never existed in fact have been no less violent in their scripting of identity than those that practise domination in the name of development . . . ' Manzo's pertinent example is that of 'traditional Zulu' leader Gatsha Buthelezi in South Africa, and there are many other examples. The crude anti-developmentalism of Pol Pot's Cambodia is a case in point (Evans and Rowley 1990; Kiernan 1985).

These points show the weaknesses of a simplistic opposition to the idea of development. Nevertheless, as the figures at the start of this chapter make clear, *actually existing development*[5] has been a profoundly contradictory phenomenon, and has in many ways failed to meet the expectations that were made for it. Development in practice, even when it has the best of intentions, shows the strength of this argument. Many development experiences have simply marginalised large sections of the population, and forced them to make a living through insecure or criminal activity. Even the most 'successful' industrial developers such as Taiwan and South Korea have had to bear an enormous cost to the environment, and have relied in part on the systematic suppression of a far from compliant work-force (Bello and Rosenfeld 1992). In a crude caricature of a supposed 'South Korean model',[6] new industrialisers in South-East Asia have relied on extensive exploitation of female labour and massive deforestation (Bello 1994). In the aforementioned case of the Green Revolution, women subsistence farmers were generally not offered the same incentives as male cash crop farmers (Young 1993: 5). This example shows how development planners have drawn on the Western idea of a sexual division of labour in which men were regarded as the breadwinners and women as the carers in the household. Such specialisation of tasks has often not corresponded to the reality of women's lives in the Third World (or indeed the First World), but it attained the status of a self-fulfilling prophecy as 'it was to men that development planners' attention turned. Women's concerns as cultivators, processors of food, traders, wage workers and unpaid labourers did not enter the planners' model' (Young 1993: 19; see also Elson 1991).

The result of this technocratic approach has been that people in the Third World have been treated as objects of development, rather than the makers of their own history, and the implementation of development within concrete social relations has had profoundly unexpected results (Kiely 1995a). Development

> fostered a way of conceiving of social life as a technical problem, as a matter of rational decision and management to be entrusted to that group of people – the development professionals – whose specialized knowledge allegedly qualified them for the task. Instead of seeing change as a process rooted in the interpretation of each society's history and cultural tradition . . . these professionals sought to devise mechanisms and procedures to make societies fit a preexisting model that embraced the structures and functions of modernity.
>
> (Escobar 1995a: 52)

The latest technicist approach is the idea propagated by the World Bank and government development agencies that the Third World (especially sub-Saharan Africa) needs to establish a better system of 'governance' (World Bank 1992). The idea is that states in the Third World should promote the rule of law, political pluralism and administrative accountability (see Moore 1993). Institutions such as the British Ministry of Overseas Development have added that democracy is desirable in the Third World, and that it constitutes a core component of the idea of good governance. While these may be desirable goals, little is said about how these might be achieved or their precise content. In fact, the emphasis appears to continue to rest with the notion that good governance rests on the capacity of states to promote development based on market forces. According to the Bank, what is required is 'not just less government but better government – government that concentrates its efforts less on direct interventions and more on enabling others to be productive' (World Bank 1989: 5).

Such attention to democracy and governance as solutions to the crisis in the Third World shows again the problems of both the content and context of development. As Schmitz argues:

> the professional manipulation of seemingly benign words and concepts belies the deepening contradiction of a modern development paradigm which remains essentially driven by techno-scientific and economistic variables, and which rests on a prevailing global power structure that is grossly *in*equitable and *un*democratic.
>
> (Schmitz 1995: 55–6)

The West's commitment to democracy ignores how the inequalities of (global and local) capitalism restrict democracy and how neo-liberals themselves wish to restrict democracy to a simple method of choosing (limited) government (Schmitz 1995: 59; Beetham 1981). Governance therefore represents the latest round of an elitist, technocratic approach to development, with all the inequalities that this entails, and which downplays the establishment of a more egalitarian global order, and of substantive, rather than formal, democracy in the Third World.

It is not surprising, then, that actually existing development has been, and is actively being, resisted within the Third World. One of the most important ways in which it has been resisted is with the rise of 'new social movements', which view the development process 'as inimical to local tradition and livelihood . . . [and] actively affirm local identity, culture, and systems of knowledge as an integral part of their resistance' (Routledge 1995: 274). Environmental movements have been established, such as the peasant-based Chipko movement in Uttar Pradesh, India, which has resisted deforestation in the region, and the Forest People's Alliance in Brazil, which has united rubber tappers and indigenous groups to resist logging and ranching practices in the Amazon state of Acre (Sethi 1993; Routledge 1995: 276–7). Civic organisations such as the South African National Civic Organisation (SANCO) have organised on the basis of struggles over housing (Bond 1995). The implementation of structural adjustment programmes has met with widespread resistance, particularly over the question of the elimination of food subsidies, which has led to anti-IMF food riots throughout the Third World (Walton and Seddon 1994). Some 'local' movements may have even taken on a 'global' character, an issue discussed in later chapters.[7] What seems clear then is that

> the Third World, or parts of it, is now engaged in a critique of the project of modernization, not least because it is constructed as another western, colonial project. This critique is bound to be ambiguous, and it is surely made possible as much by the unevenness of economic modernization as by its successes, but it does signal the depth and plurality of politics in the contemporary world.
>
> (Corbridge 1995b: 201)

Conclusion

Development since 1945 (or earlier) has in many respects failed. The Third World was always assigned a marginal place in the international economic order, and the prescriptions for advancement were either half-hearted or completely misguided. Although advances have been made for some, the pressing issues of global inequalities and poverty remain. The implementation of actually existing development has involved environmental destruction, exploitation, state oppression and impoverishment. The *global* nature of this process has entailed the integration of more and more people into a global economy, leaving many in a position of great vulnerability. In recent years, this problem has worsened as the failures of state-led development have been (partially) replaced by global market forces. In many cases, adjustment policies have coincided with, and partly caused, absolute falls in living standards.

The failings of development have not however been passively accepted, but have met with resistance from people within the Third World (and elsewhere). These movements have focused on issues around the environment, gender, agriculture, health, human rights and so on. Some, especially those that have transcended national boundaries (such as Islamic movements), have explicitly challenged Western domination of the global system. Whatever the politics of these movements (not all of which can be considered progressive), they have forced us to rethink and reshape the idea of development.

Notes

1. The concept of development is defined in more detail in the text. Some of the critical debates around this concept – such as the question of whether *all* forms of development can be rejected as a form of cultural imperialism – are alluded to later in the chapter.

2. This brief exposition of the context in which neo-liberalism emerged as the dominant paradigm in development studies needs some qualification. In my outline, I have stressed how the debt crisis and rise of neo-liberalism in the West allowed for the resurgence of neo-liberal *ideas*. However, as should become clear in the text, it is not so much the intellectual coherence (or otherwise) that allowed for the counter-revolution, but was instead the way in which these ideas effectively articulated the interests and aspirations of transnational capital. See further, Overbeek (1993).
3. These comments should not be interpreted as a blanket case against aid, or as support for the argument that aid is simply a form of imperialism. It needs to be recognised that there are many problems with the way that aid is dispensed, but this is not a case against aid *per se* (including some 'actually existing aid'). Useful assessments of aid can be found in Riddell (1987) and on food aid, Uvin (1994).
4. There is a fierce debate as to whether the terms of trade for Third World producers have shown a secular decline. The evidence appears to show that there has not been a continued decline, but that primary good production is still more risky than the production of manufactured goods. See further, Colman and Nixson (1986: ch. 5).
5. The term 'actually existing development' is derived from Bahro's (1978) conception of 'actually existing socialism'. For related terms – 'actually existing industrialisation' and 'actually existing capitalism' – see Sutcliffe (1984) and Bernstein (1990).
6. The idea of a South Korean model has been put forward by Little (1979). It has been criticised by Hamilton (1987) and Kiely (1994).
7. The issue of the politics of new social movements is a particularly contentious one. Crude anti-development positions (Marglin 1991) tend to celebrate the local as an authentic response to the homogenising thrust of globalisation. This position represents a return to the crudest versions of dependency theory in the 1970s, namely that an anti-Western position is automatically a progressive one. In fact, not all social movements can be considered progressive, and many are quite reactionary. The point remains, however, that many of these movements arose in the context of the failure of development. For example, on Hindu fundamentalist nationalism, see Harriss (1994). More generally, see Kiely (1995b).

References

Anell, L. and Nygren, B., (1980) *The Developing Countries and the World Economic Order*, London: Methuen.

Bahro, R. (1978) *The Alternative in Eastern Europe*, London: Verso.

Beetham, D. (1981) 'Beyond liberal democracy', *The Socialist Register*, London: Merlin, pp. 190–206.

Bello, W. (1994) *People and Power in the Pacific*, London: Pluto.

Bello, W. and Rosenfeld, S. (1992) *Dragons in Distress*, London: Penguin.

Bennett, J. (1987) *The Hunger Machine*, Cambridge: Polity.

Bernstein, H. (1990) 'Agricultural "modernization" and the era of structural adjustment: observations on sub-Saharan Africa', *Journal of Peasant Studies* 18: 3–35.

Bond, P. (1995) 'Urban social movements, the housing question and development discourse in South Africa', in D. Moore and G. Schmitz (eds) *Debating Development Discourse*, London: Macmillan, pp.149–77.

Brett, E. A. (1985) *The World Economy since the War*, London: Macmillan.

Colman, D. and Nixson, F. (1986) *Economics of Change in Less Developed Countries*, London: Philip Allan.

Corbridge, S. (1994) 'Post-Marxism and post-colonialism: the needs and rights of distant strangers', in D. Booth (ed.) *Rethinking Social Development*, Harlow: Longman, pp. 90–117.

—— (1995a) 'Thinking about development', in S. Corbridge (ed.) *Development Studies: a Reader*, London: Edward Arnold, pp. 1–10.

—— (1995b) 'Colonialism, post-colonialism and the political geography of the Third World', in P. Taylor (ed.) *Political Geography of the Twentieth Century*, Chichester: John Wiley, pp. 171–205.

Corbridge, S. (ed.) (1995c) *Development Studies: a Reader*, London: Edward Arnold.

Cowen, M. and Shenton, R. (1995) 'The invention of development', in J. Crush (ed.) *Power of Development*, London: Routledge, pp. 27–43.

—— (1996) *Doctrines of Development*, London: Routledge.

Crush, J. (1995a) 'Introduction', in J. Crush (ed.) *Power of Development*, London: Routledge, pp. 1–23.

Crush, J. (ed.) (1995b) *Power of Development*, London: Routledge.

Eisenstadt, S. (1966) *Modernization: Protest and Change*, Englewood Cliffs: Prentice-Hall.

Elson, D. (ed.) (1991) *Male Bias in the Development Process*, Manchester: Manchester University Press.

Escobar, A. (1995a) *Encountering Development*, Princeton: Princeton University Press.

—— (1995b) 'Imagining a post-development era', in J. Crush (ed.) *Power of Development*, London: Routledge, pp. 211–27.

Esteva, G. (1987) 'Development', in W. Sachs (1992) (ed.) *The Development Dictionary*, London: Zed, pp. 6–25.

Evans, G. and Rowley, K. (1990) *Red Brotherhood at War*, London: Verso.

Gibson, H. and Tsakalatos, E. (1992) 'The international debt crisis: causes, consequences and solutions', in T. Hewitt, H. Johnson and D. Wield (eds) *Industrialization and Development*, Oxford: Oxford University Press, pp. 41–65.

Hamilton, C. (1987) 'Can the rest of Asia emulate the NICs?', *Third World Quarterly* 87: 1225–56.

Harriss, J. (1994) 'Between economism and postmodernism: reflections on research on "agrarian change" in India', in D. Booth (ed.) *Rethinking Social Development*, London: Longman.

Kiely, R. (1994) 'Development theory and industrialisation: beyond the impasse', *Journal of Contemporary Asia* 24: 133–60.

—— (1995a) *Sociology and Development: the Impasse and Beyond*, London: UCL Press.

—— (1995b) 'Third Worldist relativism: a new form of imperialism', *Journal of Contemporary Asia* 25: 159–78.

Kiernan, B. (1985) *How Pol Pot Came to Power*, London: Verso.

Kitching, G. (1982) *Development and Underdevelopment in Historical Perspective*, London: Methuen.

Kreuger, A. (1974) 'The political economy of the rent seeking society', *American Economic Review* 64: 291–303.

Leys, C. (1975) 'The politics of redistribution with growth', *IDS Bulletin* 7(2): 4–8.

—— (1994) 'Confronting the African tragedy', *New Left Review* 204: 33–47.

—— (1996) *The Rise and Fall of Development Theory*, London: James Currey.

Little, I. (1979) 'The experience and causes of rapid labour intensive development in Korea, Taiwan province, Hong Kong and Singapore and the possibilities of emulation', in E. Lee, *Export-led Industrialization and Development*, Geneva: ILO, pp. 23–45.

Mackintosh, M. (1990) 'Abstract markets and real needs', in H. Bernstein, M. Mackintosh, B. Crow and C. Martin (eds) *The Food Question*, London: Earthscan, pp. 43–53.

Manzo, K. (1995) 'Black consciousness and the quest for a counter-modernist movement', in J. Crush (ed.) *Power of Development*, London: Routledge, pp. 228–52.

Marglin, S. (ed.) (1991) *Dominating Knowledge*, Oxford: Clarendon.

McClelland, D. (1961) *The Achieving Society*, New York: Free Press.

McNamara, R. (1975) 'The Nairobi speech', in World Bank, *Assault on World Poverty*, Baltimore: Johns Hopkins University Press, pp. 90–8.

Moore, D. and Schmitz, G. (eds) (1995) *Debating Development Discourse*, London: Macmillan.

Moore, M. (ed.) (1993) 'Good governance', *IDS Bulletin* 24: 1–79.

Moore, W. E. (1963) *Social Change*, New Jersey: Prentice-Hall.

Mosley, P. (1994) 'Decomposing the effects of structural adjustment: the case of sub-Saharan Africa', in R. van der Hoeven and F. van der Kraaij (eds) *Structural Adjustment and Beyond in Sub-Saharan Africa*, London: James Currey, pp. 70–98.

Mosley, P., Harrigan, J. and Toye, J. (eds) (1991) *Aid and Power*, London: Routledge.

Nanda, M. (1991) 'Is modern science a western patriarchal myth?', *South Asia Bulletin* 11: 32–61.

Overbeek, H. (ed.) (1993) *Restructuring Hegemony in the Global Political Economy*, London: Routledge.

Pieterse, J. and Parekh, B. (eds) (1995) *The Decolonisation of Imagination*, London: Zed.

Rahnema, M. (1991) 'Global poverty: a pauperizing myth', *Interculture* 24(2): 4–51.

Riddell, R. (1987) *Foreign Aid Reconsidered*, London: James Currey.

Routledge, P. (1995) 'Resisting and reshaping the modern: social movements and the development process', in R. Johnston, P. Taylor and M. Watts (eds) *Geographies of Global Change*, Oxford: Blackwell, pp. 263–79.

Sachs, W. (ed.) (1992) *The Development Dictionary*, London: Zed.

Said, E. (1978) *Orientalism*, London: Penguin.

Schmitz, G. (1995) 'Democratization and demystification: deconstructing "governance" as development paradigm', in D. Moore and G. Schmitz (eds) *Debating Development Discourse*, London: Macmillan, pp. 54–90.

Sethi, H. (1993) 'Survival and struggle: ecological struggles in India', in P. Wignaraja (ed.) *New Social Movements in the South*, London: Zed, pp. 122–48.

Shiva, V. (1989) *Staying Alive: Women, Ecology and Development*, London: Zed.

—— (1991) *Biodiversity: Social and Ecological Perspectives*, London: Zed.

Sutcliffe, R. (1984) 'Industry and underdevelopment re-visited', in R. Kaplinsky (ed.) *Third World Industrialisation in the 1980s*, London: Frank Cass, pp. 121–33.
Toye, J. (1983) 'Interdependence: from Kant to Brandt', in Open University, *Third World Studies: the International Setting*, Milton Keynes: Open University Press, pp. 46–66.
—— (1987) *Dilemmas of Development*, Oxford: Blackwell.
—— (1994) 'Structural adjustment', in R. van der Hoeven and F. van der Kraaij (eds) *Structural Adjustment and Beyond in Sub-Saharan Africa*, London: James Currey, pp. 18–35.
Ullrich, O. (1992) 'Technology', in W. Sachs (ed.) *The Development Dictionary*, London: Zed.
UNDP (1995) *Human Development Report 1995*, Oxford: Oxford University Press.
Uvin, P. (1994) *The International Organization of Hunger*, Geneva: Kegan Paul.
van der Hoeven, R. and van der Kraaij, F. (eds) (1994) *Structural Adjustment and Beyond in Sub-Saharan Africa*, London: James Currey.
Wade, R. (1990) *Governing the Market*, Princeton: Princeton University Press.
Walton, J. and Seddon, D., (eds) (1994) *Free Markets and Food Riots*, Oxford: Blackwell.
Watts, M. (1995) '"A new deal in emotions": theory and practice and the crisis of development', in J. Crush (ed.) *Power of Development*, London: Routledge, pp. 44–62.
World Bank (1975) *Rural Development*, Washington: World Bank.
—— (1984) *Towards Sustained Development in Sub-Saharan Africa*, Washington: World Bank.
—— (1989) *Sub-Saharan Africa: From Crisis to Sustainable Growth*, Washington: World Bank.
—— (1992) *Governance and Development*, Washington: World Bank.
—— (1993) *The East Asian Miracle*, Washington: World Bank.
—— (1994) 'Infrastructure for development', in S. Corbridge (1995c) *Development Studies: a Reader*, London: Edward Arnold, pp. 393–400.
Young, K. (1993) *Planning Development with Women*, London: Macmillan.

chapter 2

Transnational companies, global capital and the Third World

Ray Kiely

chapter 2

One of the most visible features of the process of globalisation is the rise of giant companies which operate in and influence national economies throughout the world. By the early 1990s, there were around 37,000 transnational corporations (TNCs) controlling over 200,000 foreign affiliates worldwide, generating sales of more than $4.8 trillion (UNCTAD 1994: 86). The combined sales of the world's largest 350 TNCs totalled nearly one-third of the combined GNPs of the 'advanced' capitalist countries (*New Internationalist* 1993: 18). According to some commentators (Ohmae 1991), these institutions now wield such power in the world economy that we can no longer talk about 'national economies', and instead need to focus on how these institutions undermine national sovereignty and invest wherever they like in a footloose manner.

This chapter assesses the claims that we now live in a global economy in which transnational companies have the power to by-pass particular nation-states, and especially weaker states in the Third World. It does so by examining the nature and direction of global capital flows and the influence of transnational companies in the world economy, and within the Third World. The first section (pp. 46–9) examines the 'globalisation of production thesis'. The second section (pp. 49–55) looks more critically at this thesis through an extended investigation of the nature and direction of capital flows, with particular reference to manufacturing, and the role of the TNC in promoting these flows. This section suggests that the globalisation of production thesis exaggerates the degree of (some) capital mobility in the world economy, and that this has important (and largely negative) implications for much of the so-called Third World today. The section concludes (pp. 55–62) by arguing that the movement of global capital is not completely unmanaged. Crucial here are the activities of 'local actors' who may regulate global economic forces – the 'local' reacts to and influences the 'global'. This issue is taken up in the third section (pp. 55–62) through a discussion of the developmental impact of TNCs in the Third World, where I examine their influence on both production and consumption.

TNCs and globalisation

A transnational corporation can be defined as a company that operates in more than one country. Such operations entail the setting up of productive activities, although production in this context may refer to services and finance, as well as factories, mines and plantations. The main focus of this chapter is on TNCs involved in industrial production, but some comparisons will be made with the service sector (including finance), which has undergone rapid expansion in recent years (Allen 1995: 72).

Although TNCs have operated since the nineteenth century, they have grown in significance since the 1950s, particularly in manufacturing. US-based transnationals responded to the particular conditions that existed within and beyond the domestic economy: market saturation in some sectors; a developed international communications and transporation system; and a growing economic challenge to the United States from Europe and Japan. It was against this background that US companies adopted a strategy based on an 'outward thrust to establish sales production and bases in foreign territories' (Hymer 1982: 136). European companies followed suit, and there have been new waves of expansion since, such as the increase in direct foreign investment by Japanese companies from the 1970s onwards.

By the early 1990s, there was a small number of giant TNCs whose assets were comparable to at least some developing countries' GNPs. In 1993, the largest TNC (in terms of foreign assets), Royal Dutch Shell, had total assets of $100.8 billion, $69 billion of which was foreign; General Motors, the largest motor vehicle company, had total assets of $167 billion ($36 billion foreign); and General Electric, the largest electronic company, had total assets of $251 billion ($31 billion foreign). For the same year Exxon, General Motors, Ford, Mitsubishi and Mitsui all had global sales figures totalling over $100 billion (UNCTAD 1995: 14).

But it is not just the size of these companies which is significant. According to some commentators (Ohmae 1991), the growth of these TNCs represents the development of a qualitative shift in the world economy. An *international* economy, based on trade between nations, has existed since at least the seventeenth century. In this system, the nation-state continues to be the core unit of analysis because it governs trading relations between countries – for example, through controls over the movement of capital and labour. A *global* economy, on the other hand, 'is one in which the stress is placed upon the erosion of national barriers and the movement of economic activities across national boundaries' (Allen 1995: 59). International trade remains important, but the key factor is now said to be investment flows by TNCs. These flows are said to be so mobile that national boundaries can effectively be ignored, and so economic activity can easily be shifted from one part of the world to another. In addition, TNC influence over consumption is so great that markets have now taken on a global character.

The (alleged) development of this global economy has enormous implications for the so-called Third World. Pessimists (Frobel *et al.* 1980; Peet 1986) have attempted to explain the emergence of a new international division of labour in terms of the globalisation of production. From the late 1960s, companies operating in the West and Japan faced a profit squeeze. This was caused by an increase in labour costs, which in turn was a result of trade unions winning higher wages for their workers. The response of these companies was to relocate to selected parts of the Third World, where labour costs were lower. This relocation was facilitated by the development of transport and communications which enabled companies to shift out production tasks and bring back finished products relatively quickly. It was primarily for these reasons that the newly industrialising countries of East Asia (South Korea, Taiwan, Hong Kong and Singapore) had emerged (Frank 1981).

The implications of this globalisation of production for the Third World are said to be unfavourable. Although mobile capital may take advantage of cheaper labour in the periphery, and thereby promote industrialisation, the character of this industrial development is not very desirable. It is based on low-value production and the 'super-exploitation' of Third World workers, as states lower standards (wages, conditions, and so on) in order to attract foreign investment. Special areas, often called export-processing zones (EPZs), have been established in parts of the Third World in order to attract foreign capital. Such attractions include low wages, minimal regulations and tax holidays. The result has been the growth of industrial employment but at the cost of low wages and poor working conditions, and in which the capacity of Third World states to regulate transnational corporate behaviour is severely undermined.

For the advanced capitalist states, the results of globalisation are similarly unfavourable. Capital has relocated to lower-cost areas, which has led to deindustrialisation and unemployment (Peet 1986). Thus about 30 per cent of the US trade deficit with East Asia (including Japan) is due to US firms making goods in that region and selling them in the American market. In the mid-1980s, US firms accounted for the same proportion of world exports as in the mid-1960s (18 per cent), but the US territorial economy's share of world exports fell by one-quarter over the same period (Agnew 1995: 230). The clear implication is that in order to compete in the global economy, First World countries will have to downgrade their economies in terms of labour costs and conditions. Capital mobility and free trade, promoting low-wage exploitation in the Third World and sustained unemployment in the First, is, for the pessimists, a likely scenario for the foreseeable future (Lang and Hines 1993; Williams *et al.* 1995).

More optimistic theorists of globalisation, on the other hand, argue that the growth of 'stateless' corporations is desirable because it undermines the capacity of inefficient states to regulate and distort the efficiency of market forces. Everyone can benefit from the opportunities of global market forces if they are prepared to shed the burdens of counter-productive government intervention (*The Economist* 1994; Ohmae 1991). This viewpoint is shared by some politicians – the Conservative government in Britain, for instance, opposed the European Union's Social Chapter on the grounds that it would undermine European and particularly British competitiveness.

Although the optimists and pessimists disagree about the implications, they are at one in their basic claim: the increase in the power and scope of TNCs has given rise to a new, global economy in which the capacity of nation-states to regulate their economies has been undermined by the 'hyper-mobility' of capital (Harris 1994). For the Third World, this has led to new types of production, most notably the rise of industry. The optimists believe that this is the way forward for the Third World as a whole, while the pessimists argue that the state is at the mercy of powerful companies that can easily relocate to lower-cost areas if they so wish.

What can be made of these claims? Certainly there is evidence to back them up, but so too is there evidence that undermines them. In the next section (pp. 49–55), I will critically examine the claims made for a global economy, and suggest that while they point to one *tendency*, there are important *counter tendencies*, which suggest the need for a more balanced approach, which takes account of the complexities and specificities operating in the world today. In so doing, I challenge the claims of *both* optimists and pessimists. I do so by first examining the nature and direction of global capital flows. This evidence is then used to suggest that the claim that we are now living in an unregulated global economy is exaggerated.

Explaining global capital flows

Global capital flows can be divided into manufacturing and services. In the service sector there has been an increase in direct foreign investment in recent years, which reflects the fact that there are fewer restrictions on foreign investment in this sector, and the fact that services account for an increasing proportion of economic activity in the developed countries. In service industries such as catering and tourism, it is difficult to separate the consumption of the service from its production and so there is an obvious incentive to invest abroad – imported (finished) hamburgers would be rather cold and would cease to be fast food! In the financial sector, it is the case that money can easily be moved around the globe and therefore hinder national regulation. It is this sector above all others which can most accurately be described as globalised (Allen 1995: 72–3; but see also the Introduction, pp. 10–11). My principal focus here though is on industrial capital. This remains statistically the most significant in terms of foreign investment flows, and a strong manufacturing base remains vital to the competitiveness of any economy, a point I return to below. Nevertheless, some of the data presented in this section refers to foreign investment in general, and not just to the sphere of manufacturing.

For the period from 1991 to 1993, developing countries as a whole received around 31 per cent of the total global stock of direct foreign investment (DFI). This investment was in turn highly concentrated in particular regions – Africa received 1.7 per cent of the world total; Latin America and the Caribbean 9.8 per cent; West Asia 0.8 per cent; and East, South and South-East Asia 18.8 per cent (UNCTAD 1995: 12). Moreover, just ten nations accounted for 68 per cent of DFI in the Third World in the early 1990s (*New Internationalist* 1993: 19). These were Singapore, Mexico, Brazil, China, Hong Kong, Malaysia, Egypt, Argentina, Thailand and Taiwan. Hirst and Thompson (1996: 67–8) have estimated that in the early 1990s around 28 per cent of the world's population received 91 per cent of global DFI and so around 72 per cent of the population received 8 per cent of global DFI.[1] Such figures suggest that the developing world refers to an increasingly diverse set of countries, and that it may no longer make sense to talk about a unified Third World at all. On the other hand, global hierarchies still exist and have actually intensified in the global era, as I show below (pp. 49–55).

Most foreign investment is therefore concentrated in the already developed countries, or in the more successful developers in the (former) Third World. There is a general trend in which the share of direct foreign investment made by First World TNCs in the Third World has actually *fallen* in recent years – from 27 per cent (1975) to 19 per cent (1984) for West Germany, from 73 per cent to 52 per cent for Japan, and from 19 to 16 per cent for Britain (Jenkins 1992: 35).

Moreover, despite the growth of direct foreign investment, most TNCs continue to concentrate their operations in their home region. This is the case for motor vehicle companies like General Motors and Ford, for electronics companies such as Matsushita and General Electric, and even for the tobacco company BAT (UNCTAD 1995: 14). Although the picture is less clear in other sectors (such as food and computers) where combined foreign assets are greater than domestic assets, 'the world is still one of national economies and protected markets' (Allen 1995: 67). This may vary according to the size of the domestic market (for example, Nestlé's sales in the Swiss market constitute a small proportion of its global sales), but domestic and regional markets remain central. IBM's sales in the US computer market constitute almost two-fifths of their total global sales, while Unilever's sales of food and household products in Britain and the Netherlands are more than 50 per cent of their global sales (UNCTAD 1995: 14). Of the top 100 firms in the world in 1993, only 18 kept the majority of their assets abroad (Wallace 1996: 24). The proportion of sales in the home region of TNCs in the 'advanced' countries remains very high or has actually increased in recent years: German manufacturing TNCs sold 75 per cent of their goods in their home region in 1993, compared to 72 per cent in 1987; Japan sold 75 per cent, compared to 64 per cent in 1987; while US manufacturing TNCs sold 67 per cent, compared to 70 per cent in 1987. In services, there has been a percentage decline in domestic sales, but these are still higher (Japan 77 per cent, UK 77 per cent, US 79 per cent) than foreign sales. In terms of assets, the figures are similarly striking – 93 per cent of Japanese manufacturing TNC assets were held domestically in 1992–3; in the United States the figure was 73 per cent, and even in the case of British TNCs, which have a reputation for investing abroad more than at home, the figure was 62 per cent (Hirst and Thompson 1996: 96). Moreover, it should again be stressed that most of the assets and sales of the TNCs outside of their home country is in other First World countries, and not in the Third World.

Global trade patterns also suggest that the globalisation thesis is exaggerated. The share of exports as a proportion of the GDP of 'advanced' capitalist countries rose steadily from the 1950s to the 1980s, but by the late 1980s this proportion was still lower than it had been in the period immediately before the First World War. The share of Africa, Asia and Latin America in world trade has actually gradually declined: Latin America's share of world exports fell from 12.4 per cent in 1950s to 3.9 per cent in 1990; Asia's increased from 13 per cent in 1950 to 17.8 per cent in 1980, but fell back to 14 per cent in 1990; while Africa's share declined from 5.2 per cent in 1950 to 1.9 per cent in 1990. Although the developing world as a whole has increased its global share of exports, this improvement is largely accounted for by the export performance of the four original Asian tigers, who produce around half of the total manufacturing exports originating from the Third World (Glyn and Sutcliffe 1992: 79, 90–1).

Although institutions such as the General Agreement on Tariffs and Trade

(GATT) and the newly established (in 1995) World Trade Organisation represent important movements towards the establishment of a truly global economy, there are important counter tendencies. First, significant barriers remain in place, such as voluntary agreements between countries (which limit trading activity), anti-dumping protection (selling goods at below cost) restrictions, import quotas on selected items (such as clothing) and national subsidies (that is, indirect protectionism). These restraints particularly affect the developing world (Page 1994). Second, given the evidence and explanation for capital flows cited above, it seems unlikely that enhanced capital mobility would increase the proportion of global capital flows directed towards the Third World. Third, Article 26 of the GATT permitted free trade areas on the basis of a total dismantling of trade barriers between members of the regional groupings. The European Union and the North American Free Trade Agreement (NAFTA), among others, could be said to represent an important step up the ladder from national to regional and eventually global free market economy. However, the picture is more complex than this. Within free trade areas there remain important barriers to trade which relate to health regulations, government support and national production standards; in March 1996, for example, member states of the European Union (and beyond) imposed a ban on the import of British beef. Moreover, free trade areas may actually become regional protectionist groupings, taking a fortress mentality to goods and labour outside of its jurisdiction. The European Union was for instance instrumental in slowing the progress of the Uruguay Round of the GATT from 1978 to 1993 (Brook 1995: 144–5; World Trade Organisation 1995).

How strong then is the claim that we are now living in a global economy, in which capital is footloose and can easily move from one region to another, including from First to Third World? Clearly such a view exaggerates the extent to which *productive* capital has moved from First to Third World. Advocates of the globalisation thesis outlined in the first section (pp. 49–64) have argued that parts of the Third World have industrialised as capital from the First World has relocated to low-wage areas in the Third. But as I have shown, if anything, the tendency has been for capital to *concentrate* in established areas of accumulation – it has moved out of the Third World as a whole. The figures above do show that the South-East and East Asian industrialising regions have been a favoured site for foreign investment, but this cannot be the sole explanation for their industrialisation. In both South Korea and Taiwan, until recently there have been strict restrictions on foreign investment, and the major actors in their industrialisation processes have been local capital in close alliance with the state (Kiely 1997: chs. 7 and 8; Wade 1990). In Taiwan, for example, foreign firms accounted for only 5.5 per cent of capital formation in the years of rapid economic growth from 1962 to 1975 (Wade 1990: 149). The lessons of the Asian miracle economies in fact 'show the value of determined national economic management and solidaristic public policies in producing international competitiveness. This is the exact opposite of what most globalization theorists argue: success in the international economy has *national sources*' (Hirst and Thompson 1996: 114–15; see also Kiely 1994).

As regards the nature and direction of global capital flows, a number of conclusions can be drawn from the data. The mobility of productive capital is limited by the fact that investment in plant and equipment is relatively immobile. This lack of mobility increases the higher the level of capital intensity – in other words, heavier industries are far less mobile than labour-intensive ones. Thus the decline of the steel and shipbuilding industries in Britain was not a product of relocation to other parts of the world, but was instead caused by the loss of Britain's competitive position in these sectors. The second, closely related point is that there are factors which serve to grant established areas of capital accumulation competitive advantages in the world economy. Early developers enjoy a relative advantage over later developers as they secure economies of scale, or the utilisation of methods of production which enable them to produce goods more efficiently than later developers. These include the use of mass production techniques, access to cheap credit and inputs, developed infrastructures and the organisation of research and development activities. In addition, the most developed markets are concentrated in these areas (Kiely 1997: ch. 5). In practice, then, capital is largely attracted to existing areas of capital accumulation. Late developers may of course buy the most advanced technology from the First World, but this in turn leads to a whole new set of problems. Leaving aside the cost of acquiring such technology,[2] the nature of competition often leaves later developers at a disadvantage. As Amsden notes:

> Foreign competitors may be expected to introduce a stream of new innovations in productivity and quality to retaliate for their loss of market share, and it takes time for late industrializers to build a team of engineers and a work force with the capabilities to keep abreast of these advances. Even the initial process of technology transfer is fraught with problems because technology is never completely codified, and to operate it optimally in different environments requires adaptation. Moreover, the quality of human and physical infrastructure and the efficiency of suppliers vary internationally. Even the most enterprising late-industrializing companies have no control over these inputs, but their productivity levels are likely to be significantly influenced by them.
>
> (Amsden 1992: 55)

Such capital agglomeration can be said to be occurring at a global level. Thus it is not so much the hyper-mobility of capital that is important, but rather *the relative immobility of productive capital*, which is intensifying uneven development in the global economy. Established centres of capital accumulation (the European Union, North America and Pacific Asia – although there is uneven development too *within* these regions) are attracting most of the share of capital investment, while other parts of the world are increasingly being marginalised from this process. The result is an intensification of global inequalities, both at the level of nation-states and in terms of global inequalities based on class. In 1900 the gap between the richest and poorest countries was around 8:1; by the late 1980s this had increased to 36:1 (Freeman 1991: 155). In

1960 the richest 20 per cent of the world received 70 per cent of global income, while the bottom 20 per cent received 2.3 per cent; by 1989 the richest 20 per cent received 82 per cent of global income while the bottom 20 per cent received 1.4 per cent. The ratio of rich to poor income earners therefore increased in this period from 30:1 to 59:1 (*New Internationalist* 1996: 19). Hirst and Thompson are therefore right to assert that

> In this ambiguous international system the problem for poorer countries is not imperial domination or attempts to annex their resources, it is neglect and exclusion. Trade and investment flow between advanced states and a few favoured newly industrialized countries. The rest have become economically marginal.
>
> (Hirst and Thompson 1995: 419)

Having said that, the concentration of capital in certain areas is only tendential, and is subject to certain important counter tendencies. If this were not the case, then we would have no explanation for the relative decline of Britain or the rise of new areas of capital accumulation, such as the East Asian NICs. One important factor is the action of states in different 'localities' within the world economy, which can act to counteract these tendencies (see pp. 55–62). Another important factor is the existence of some sectors in which there is greater mobility for capital, and certain competitive advantages for Third World economies. This is particularly true in labour-intensive sectors such as textiles, toys and some electrical components. In these cases, fixed costs are not too high as technology is not particularly advanced. This gives capital a potentially far greater mobility and provides some cost advantages for Third World countries. This is because labour costs constitute a higher proportion of total costs in these sectors, and so cheaper labour in the periphery can act as an advantage. In so far as there was relocation by TNCs from the late 1960s, it tended to be in these sectors. In the 1980s and 1990s, relocation of selected industries continues. For instance, as wage costs have increased in Taiwan, South Korea and Hong Kong, labour-intensive industries have shifted their operations to lower-cost countries in the region such as Thailand, Indonesia and special economic zones in south-east China.

It is these sectors which most closely correspond to the theory of the globalisation of production outlined in the first section (pp. 46–9). In these cases, labour considerations are paramount. This will include not only costs but also the extent to which labour is unionised (and the character of unionisation) and, in some cases, labour skills. At the lowest end of this labour market are the so-called world market factories, employing workers for low wages and long hours in sometimes appalling conditions. Employment in these factories often takes place within the aforementioned EPZs. Employers often take advantage of cheap female labour. Low wages are 'justified' by the argument that the proper place for women is in the home, and the skills involved in production (such as sewing) are somehow natural (Pearson 1994). The growth of women's employment has led some writers (Standing 1989) to claim that there is a global feminisation of the labour force, in which employers take advantage of super-exploited cheap female labour in order to survive in a competitive global economy.

Although the growth of female labour is an important tendency in the world economy, this phenomenon should not be exaggerated. As should be clear by now, these world market factories and EPZs represent only one tendency in the world economy, and operate largely in a few labour-intensive sectors. Around 200 EPZs operating in 50 countries currently employ around 2 million workers, although this figure excludes the special economic zones (SEZs) in China (Gereffi and Hempel 1996: 22). Employment in EPZs rarely accounts for more than 5 per cent of total industrial employment within individual countries, and in many cases (such as India, the Philippines and Taiwan) less than 10 per cent of total manufactured exports originate from EPZs (Jenkins 1987: 132).

As well as the growth of EPZs, TNCs have increasingly sub-contracted out production activities to local companies in the Third World. Companies such as NIKE and The Gap do not actually own production facilities, but instead contract out such work while the company proper concentrates its activity on design and marketing (Donaghu and Barff 1990; Mitchell 1992). In these cases, higher-quality consumer products are being produced and so labour skills are equally as important as costs, and so productive activity is more likely to be contracted out to higher-tier developing countries such as South Korea (Gereffi 1994: 220–3).

One useful way of conceptualising these tendencies is to think in terms of *global commodity chains*. A commodity chain can be defined as 'a network of labor and production processes whose end result is a finished commodity' (Hopkins and Wallerstein 1986: 159). A global commodity chain is one that links such processes at a global level. As Gereffi and Korzeniewicz (1994: 1) state: 'Capitalism today thus entails the detailed disaggregation of stages of production and consumption across national boundaries, under the organizational structure of densely networked firms or enterprises.' Gereffi usefully distinguishes two kinds of global commodity chains. The first, producer commodity chains, broadly coincide with the capital-intensive sectors outlined above. This is where the site of production of final product is less mobile and tends to agglomerate within established areas of accumulation. In these cases, 'manufacturers making advanced products like aircraft, automobiles and computer systems are the key economic agents in these producer-driven chains not only in terms of their earnings, but also in their ability to exert control over backward linkages with raw material and component suppliers, as well as forward linkages into retailing' (Gereffi 1994: 219). The second, buyer commodity chains, are characterised by more mobility and production is more labour intensive and therefore more likely to take place in the Third World. However,

> these same industries are also design- and marketing-intensive, which means that there are high barriers to entry at the brand-name merchandising and retail levels where companies invest considerable sums in product development, advertising and computerized store networks to

> create and sell these items. Therefore, whereas producer-driven commodity chains are controlled by core firms at the point of production, control over buyer-driven commodity chains is exercised at the point of consumption.
> (Gereffi 1994: 219)

Thus, to sum up, the following points can be made. First, the case for the existence of a hyper-mobile global capital is an exaggeration. In fact, much productive capital investment is relatively immobile, and this is a major reason why much of the Third World is marginalised from global capital flows. In the case of labour-intensive industries, there is greater capital mobility and better investment prospects for Third World nations, but there remain important (marketing) barriers to entry in these sectors. Second, it follows from the first point that the tendency towards globalisation has intensified, rather than alleviated, uneven development in the world system. The move towards 'global market forces' has increased global hierarchies and inequalities, as some dominant regions grow while others remain relatively marginalised. Economic globalisation therefore does not imply economic homogenisation, but has led to an increase in differentiation. Thus, contrary to the claims of optimistic globalisation theorists, the tendency is for global market forces to intensify rather than alleviate inequalities, precisely because the world market is not a level playing field. The third, and more optimistic, point is that local states are not irrelevant, and are far from being the passive victims of a footloose, hyper-mobile global capital. Although there may have been moves towards a lowering of work standards in labour-intensive sectors, states can also play a crucial role in upgrading those capital-intensive and high-technology sectors which are more reliant on skills, and can afford to pay higher wages in return for high productivity. The pessimistic globalisation thesis outlined above fails to appreciate this crucial point, as the decline in industrial employment was not a product of relocation by mobile capital from First to Third World, but was actually caused by an increase in productivity through technological advance, and a consequent decline in the demand for industrial labour (Rowthorn and Wells 1987: ch. 1). Such advances were often instigated by active state intervention, which often promoted specific industries through credit allocation, state planning, subsidies and forced investment in particular sectors (on Japan, see Johnson 1982). Although states in the Third World may be weaker in their attempts to regulate the control of TNCs, they are not passive. It is to a consideration of the role of TNCs in the Third World that I now turn.

TNCs and the Third World

As well as the issue of global capital flows, the debates concerning the impact of direct foreign investment within particular Third World nation-states deserve some attention. Put differently, how does one explain the interaction of the *global* (transnational investment) and the *local* (states, classes, consumers and so on)? The following section examines this question through an investigation of two issues. First, the economic impact of TNCs is examined, particularly their impact on *production* in the Third World; and second, their impact is examined through an account of the relationship between TNCs and *consumption* patterns. This second question involves some further examination of the economic impact of TNCs, but also some consideration of wider concerns, including an introduction to some of the debates around the issue of cultural imperialism (taken up further in Chapter 7). In addressing these matters, some contrasts are made between my own position and that of other writers on TNCs, who often fall into one of two over-simplistic positions, based on outright apology for, or condemnation of, the activities of transnational companies.

TNCs and local production

TNCs invest in Third World countries for a variety of reasons. These include gaining access to domestic markets, to specific raw materials and cheap labour and to avoid state regulations which may be stronger in the First World. A lot of writing on TNC investment has tended to make sweeping generalisations concerning their developmental effects in the Third World – for example, as we have seen, some writers have wrongly argued that Third World industrialisation from the 1970s was solely a product of TNC relocation in search of cheap labour. The reality was that in the most successful NICs (Singapore apart), it was local capital, in alliance with the state, that led the process of sustained industrial development (Kiely 1994). In examining the effects of TNCs, then, we should bear in mind that we can only describe some tendencies and counter tendencies, which will operate in different ways in specific situations. This should become clearer if we examine the developmental effects of TNCs in five specific areas: capital stock and income; linkages; technology; employment and labour; and political stability.

Capital stock and income

Apologists for TNCs argue that TNC investment increases the capital stock and therefore the income within a country, including foreign currency earnings (May 1975). Critics argue that such investment does not necessarily increase the income of a particular country as capital outflow may exceed inflow (Frank 1972: ch. 8). Moreover, TNCs may *transfer price*. This is where two subsidiaries of the same parent company trade with each other and so manipulate the prices paid for such goods. This practice enables companies to evade tax payments to nation-states – profits may be declared in low-tax countries, and thereby effectively hidden in high-tax countries. Some states are thereby deprived of revenue through the actions of tax-evading TNCs (Murray 1981).

What can be made of these conflicting claims? The apologist case is at least exaggerated – as shown above, significant numbers of TNCs sub-contract to local producers rather than investing directly. TNCs in the food and drink sectors take advantage of their brand name and 'secret ingredients', and license these out to local capitalists without making any direct investment. Even in the case of direct investment, capital may be raised within the country in which they invest (Jenkins 1987: 96). Having said that, not all of the funds are raised locally, and even if they are, TNCs are at least putting such funds to productive purposes. Moreover, the critics' argument that outflows exceed inflows is very unsatisfactory, because it fails to account for the fact that the money that stays within a country can be used for productive purposes, and so act as a stimulant to further economic activity (Warren 1973). Even more important, the critics' case is full of logical inconsistencies. As Jenkins explains:

> critical accounts of the 'drain of surplus' are unsatisfactory in that they remain at the level of appearances, being content to show the existence of a net outflow of capital without providing an adequate theoretical explanation. This gives rise to an apparent paradox wherein TNCs are seen as exploiting less developed countries by making excessively high profits, while at the same time they repatriate rather than reinvest profits despite such high profit rates.
>
> (Jenkins 1987: 98)

Taken to its logical conclusion, the drain of surplus thesis would have to argue that those countries with the most TNC investment will be the poorest countries, while those with the least investment will be the richest – clearly an absurd contention.

The explanation for why capital outflows often exceed inflows in the Third World must lie with the potential for long-term profitable opportunities existing in the world economy. As the second section above made clear, capital has tended to concentrate in specific regions where long-term profitable opportunities are greater, and is not dispersed evenly throughout the world. Capital outflows exceeding inflows should thus be seen as a *symptom* rather than a *cause* of uneven development (Jenkins 1987: ch. 5). Furthermore, such outflows are not only caused by TNCs, but by local elites also 'wanting a piece of the action'.

Finally, transfer pricing is undoubtedly a common practice among TNCs but, given the fact that taxation rates are often lower in the Third World than in the First, it is unclear why the Third World should be deliberately singled out for such practices (Corbridge 1986: 172).

Linkages

The extent to which TNCs open up or close off linkages to the rest of the economy depends largely on the state of the local economy and the sector in which TNCs operate. TNCs have been notorious for creating enclave-type economies in mining and extraction industries such as bauxite (Girvan 1976), and in the labour-intensive industries in export-processing zones (Jenkins 1987: 111). In these sectors, TNCs have largely imported their requirements and so forward and backward linkages to the local economy have been limited. In the case of US industries operating in Latin America and the Caribbean, US tariff laws have actively discouraged the purchasing of locally made inputs because only US-made components are exempt from import duties when the finished product is shipped back to the United States (Gereffi and Hempel 1996: 22).

On the other hand, one cannot overgeneralise from the experience of labour-intensive manufacturing or extractive industries. In heavier industries such as automobiles in Brazil, TNC subsidiaries do purchase from local suppliers and so some linkages are generated. TNCs may have a tendency to import greater amounts than local capital, but this is less a product of the nationality of capital and more a question of the sector in which foreign or local capital operates (Lall and Streeten 1977). Indeed, when operating in similar sectors, it is far from clear that local capital would behave any differently from foreign capital – if viable local suppliers exist, then capital will utilise them. When they do not, then local or foreign capital will import from abroad (Lall 1978).

Technology

Apologists for TNCs argue that they introduce the most advanced techniques to the Third World, and therefore provide them with the means to expand output and incomes and compete effectively in the world economy (Vernon 1973). Moreover, such technology is acquired relatively cheaply as the research and development expenses that promoted such innovation were met by the First World (Warren 1973: 30–1; Emmanuel 1982). Critics point to the restrictions placed on purchasers, such as tying buyers to certain suppliers, or restrictions on exports. They also rightly point out that the acquisition of technology is no guarantee that it can be used optimally, and that this is most likely to occur through the development of an indigenous research and development capacity. TNCs, on the other hand, tend to concentrate such activities within their home base, so that leakages to potential competitors can be avoided (Lall 1993a). Having said that, there is little evidence to suggest that *on its own* local capital will be more likely to develop such a capacity. Indeed, in the cases of Taiwan and South Korea, successful research and development facilities were generated by the activities of a strongly interventionist state (Kiely 1997: ch. 8; Singh 1994: 22).

The extent to which imported technology may be appropriate for the Third World economy has also been questioned. For example, capital-intensive technology in the 1950s import-substituting industries in labour-abundant Latin America had a limited impact on job creation (Hymer 1982). The apologist case for importing technology is weakened by its implicit belief that the introduction of the most advanced technology will automatically serve people's needs, when in fact it may simply displace workers from jobs or at least not create many new employment opportunities. On the other hand, the appropriate technology critique is similarly naive in its assumption that appropriate technology could be introduced in the absence of TNC activity. As in the case of linkages outlined above, there is little reason to believe that local capital operating in the same sector as foreign capital will behave any differently from TNCs in their choice of techniques. As Kiely (1997: ch. 5) has argued, 'questions of appropriate technology do not depend on the "nationality" of that particular technology, but rather on the place of that technology within capitalist social relations'. The competitive accumulation of capitals forces firms to introduce the most advanced technologies in order to lower production costs, increase profits and ultimately stay in business. States in the Third World *may* regulate these activities – with varying degrees of success – but the question then becomes one of state–capital relations, rather than the impact of TNCs on developing countries.

Employment and labour

Apologists argue that TNCs create jobs, both through direct investment in the Third World, and through positive spin-off effects such as higher demand through wages and use of local suppliers. Wages for workers employed in TNCs are often higher than wages for workers employed by local capital too. Some qualifications need to be made however. TNC investment may lead to the displacement of workers previously employed by local capital put out of business through the operations of TNCs. Higher wages paid by TNCs are often more than offset by higher productivities in these sectors, and the wages are still likely to be less than those paid to workers in similar jobs in the First World (Vaitsos 1976). Also in some labour-intensive sectors such as textiles and agriculture, TNCs pay low wages (though usually not lower than local capital) and work conditions can be appalling. This is especially true of the world market factories in export-processing zones and plantation production in agriculture.

The picture then for TNCs and employment is highly variable. Few (but still some) jobs are likely to be created in sectors using capital-intensive or high technology, while labour-intensive sectors are likely to create more jobs but at the cost of poor working conditions. Once again, local capital appears to operate in a similar way to TNCs, and is equally as likely to invest in capital-intensive equipment, or make use of world market factories or plantation agriculture.

Political stability

Apologists argue that by promoting economic growth, TNCs also promote long-term political stability. Critics cite examples whereby TNCs have actively played a part in undermining governments in the Third World – the activities of ITT in Chile in 1973 and of food TNCs in Central America throughout the twentieth century are the two best-known examples. In addition, TNCs may take advantage of the Third World's willingness to attract foreign investment and so locate to 'pollution havens', where environmental controls are less strict (Michalowski and Kramer 1987). This may have disastrous consequences, as was the case with the chemical leak from a Union Carbide pesticides factory in

Bhopal, India, which led to the deaths of over 2,000 people and the maiming of a further 20,000 (Smith 1992: 286).

It would be a mistake however to overgeneralise from these undoubtedly powerful examples. States in the Third World are not simply passive victims of the activities of TNCs, and some have quite successfully regulated the activities of foreign capital for their own developmental ends. In the 1970s, for example, a wave of nationalisations throughout the Third World forced TNCs into making some concessions to local (state) capital, although in many cases the former retained significant power through control of international distribution and marketing networks (Elson 1988: 268–72). Perhaps most dramatically, states in East Asia such as Taiwan and South Korea have imposed strict controls over foreign investment (although there has been considerable liberalisation since the 1980s). These developmental states have successfully 'guided the market' (Wade 1990) so that capital has to some degree faced the discipline of state regulation. Such regulation has included state allocation of credit which has favoured specific sectors, state-led research and development and limitations on capital outflow (Amsden 1989). Such practices have particularly affected local capital, but East Asian states have also regulated TNC activity, through for example not allowing foreign investment in some sectors (foreign loans were preferred), but drawing on some of the more dynamic elements in others, through joint ventures between local and foreign capital. This policy was an attempt to control the negative aspects of foreign investment, while constructively drawing on the more positive ones such as technology transfer (Bello and Rosenfeld 1992: 54–5).

More recently, the implementation of structural adjustment policies (discussed in Chapter 1) has undermined the capacity of local states to regulate the behaviour of TNCs, and economic activity more generally. However, even in this case the role of the state and other local actors is far from irrelevant. The implementation of adjustment policies has involved complex negotiation between the international financial institutions and local states, and local resistance has in some cases limited the full implementation of austere economic policies (Mosley *et al.* 1991; Walton and Seddon 1994). Moreover, the economic liberalisation policies associated with adjustment do not entail an elimination of the role of the state, but a change in its regulatory activity (UNCTAD 1994: 148). Indeed, the promotion of a free market economy rests on the existence of a strong state, a point that even the World Bank appears to accept (World Bank 1992).

This brief discussion shows the problems of making easy generalisations about the impact on production of TNCs operating in the Third World. Whilst it would be an optimistic fallacy to claim that TNC investment unproblematically expands output and income in the periphery, so too is it wrong to claim that it is the primary factor in the 'underdevelopment' of the Third World. Put differently, it would be one-sided to argue that global actors (in this case transnational corporations) have made local actors (particularly the state) irrelevant. For example, the more export oriented nature of foreign subsidiaries in South Korea is less a product of higher export intensity by TNCs, and more a product of the fact that the state has reserved the market for domestic firms (Jenkins 1993: 127). TNCs do not automatically dominate local actors, but, by the same token, the two do not necessarily conflict with each other. Both foreign capital and the state may for example have mutual interests in supporting the expansion of capitalist industrialisation within a particular country, although this shared interest may be mediated by conflicts over the distribution of the benefits – the 'triple alliance' of state, local and foreign capital in Brazil in the 1960s and 1970s is one example (Evans 1979). In terms of TNCs and Third World states, then, the most important issue is not the 'nationality' of capital, but the wider social and political context in which TNCs (and other capitals) operate.

This particular configuration of 'global' and 'local' factors can also be seen in the case of the relationship between TNCs and consumption patterns in the Third World.

TNCs and local consumption

Critics of TNCs in the Third World argue that their marketing strength enables them to manipulate 'choices', so that consumers buy particular brand names. Schiller (1979: 23) argues that the effect of TNC advertising 'has been to create audiences whose loyalties are tied to brand named products and whose understanding of social reality is mediated through a scale of commodity satisfaction'. There is some (limited) evidence that TNCs spend more on advertising than local companies (Jenkins 1988), and that some TNCs are responsible for promoting 'inappropriate products'. From 1970 onwards Nestlé was accused of promoting the sale of powdered milk for babies in areas where potential customers did not have access to clean water, which led to serious health consequences (or worse) for babies (Muller 1982). Some TNC drug companies have flooded Third World markets with a variety of high-price brand names, rather than promote one basic, cheap generic drug. Some products banned in the First World have been sold in the Third, and others have been used as a kind of laboratory test in the Third World before being made available in the First (Gereffi 1983). Such examples have led critics to argue that TNCs are agents of cultural imperialism, which 'claims that authentic, traditional and local culture in many parts of the world is being battered out of existence by the indiscriminate dumping of large quantities of slick commercial and media products, mainly from the United States' (Tunstall 1977: 57).

However, at least three qualifications need to be made to this claim. First, the evidence that TNCs advertise more than local companies is far from clear cut – in the pharmaceutical industry for instance (one of the worst offenders according to the cultural imperialism thesis), local companies (where they are reasonably strong, as in Argentina) actually advertise more heavily than TNCs (Jenkins 1993: 118). Second, even in the relatively straightforward cases of inappropriate products outlined above, the problem is not one of TNCs simply imposing their will on passive Third World countries. One also needs to address the failure of local states to regulate the behaviour of TNCs. Third, and perhaps most important, is the question of how far we can generalise from these examples and condemn TNC behaviour across the board. In other words, what is wrong with Third World societies being sold Western products? Who is to decide what is good for the people of the Third World? Tunstall's reference to the authenticity of Third World cultures as opposed to the slick commercial products of the West itself betrays a form of cultural imperialism, based on the romanticisation of 'other' cultures (Tomlinson 1991: 120–1).

These critical comments should not be taken as an endorsement of the idea that consumers are sovereign in the market place, or that we should simply celebrate popular consumption (see McGuigan 1992; and Chapter 7 of this volume). Consumption patterns are closely tied to income, which is particularly unequally distributed within many Third World countries. TNC entry into Third World markets does have an impact on the production and marketing strategies of local firms, which often entail increased capital intensity and far more aggressive marketing techniques (Jenkins 1993: 125). In this way, TNCs play a vital role in promoting the *tendency* towards a global standardisation of tastes.

However, TNCs cannot themselves take all the blame for such an occurrence. Consumption operates in a particular *social context*, of which TNCs are but one part. It is a context in which consumers have very limited control over their own lives, and this impinges on what is actually consumed – fast food is eaten not for its intrinsic quality, but for its convenience; a car is bought because it 'is impossible to live without one'. If we regard consumption as a social activity, then the critique of cultural imperialism becomes less a critique of TNCs in themselves, and more one of the social context in which they operate. As Tomlinson argues:

> To grasp the order of 'blame' in cultural domination we have to think of capitalism as something wider than the practices of individual capitalist organisations, however large and powerful. This wider view of capitalism as one of the key autonomized institutions of modernity represents it as something within which the routine practices both of ordinary people and of individual capitalist organisations are locked. It is in this wide sense that we can speak of a 'culture' of capitalism. But here the notion of 'blame' is more problematic: it is not individual practices we are blaming, but a contextualising structure: capitalism not just as economic practices but as the *central (dominant) position of economic practices* within the social ordering of collective existence.
>
> (Tomlinson 1991: 168, emphasis in original)

Conclusion

Transnational corporations represent one important factor in the globalisation of capital. Both the size of transnational companies and foreign investment flows across countries have increased substantially, and this has affected both production structures and consumption patterns in the Third World. Nevertheless, both the optimistic and pessimistic advocates of the extreme globalisation thesis, which suggest that capital is hyper-mobile and the nation-state largely irrelevant, are mistaken. The contemporary world economy is not witnessing footloose capital investment dispersing throughout the world, so much as a *concentration* of capital, investing in certain areas and largely marginalising others. The level playing field of the optimistic globalisation thesis, in which all participants in the world economy can equally benefit from global market forces, is a myth. As Lall (1993c: 6) points out, '[e]conomic liberalization, while making TNC entry more welcome and procedures less obstructive, is likely to increase the concentration of investment flows towards countries that offer a large internal market, internationally competitive skills, support systems and infrastructures'. Thus the concentration of capital flows in established areas of accumulation is likely to increase with liberalisation. For similar reasons, productive capital is unlikely to take advantage of low wage costs in the developing world, except in the most labour-intensive sectors such as clothing. First World deindustrialisation, in so far as it is likely to continue, will largely remain a product of technological innovation rather than TNC relocation to the Third World. None of this suggests that the developmental prospects for Third World societies are impossible. The concentration of capital flows should be regarded as a tendency operating at a global level, but no more than that. Such tendencies can be counteracted by the activities of local actors, such as local capital, the state or indeed political and social movements. Global capital flows can be affected by 'local action'. Thus the movement of capital into the NICs is partly a product of earlier successful state-guided management of local capital, particularly in East Asia. Such global–local relations also operate when TNCs invest within particular Third World economies. The developmental effects of TNC investment are neither uniformly bad nor good, but are in fact the product of a number of contingent factors, such as the sector in which the TNC operates, the character of local capital, the role of the state, and specific interests within a country. It is this particular social context, which is simultaneously both global and local, that is crucial in determining the nature of development within a particular country.

Notes

1. These figures are based at the level of nation-states rather than people in nation-states. In the real world not all people within a country will benefit from investment – indeed, many of the major recipients of DFI (First as well as Third World) have high rates of unemployment. The figure of 28 per cent is however arrived at by including only those areas of China (eight coastal provinces plus Beijing province) that are favoured recipients of direct foreign investment.
2. There is an extensive literature on the costs of importing technology. The pessimists point to the expense and the difficulties of effective local utilisation, while optimists

argue that, given that R&D expenses are met from abroad, such technology is acquired relatively cheaply by importers. For summaries which attempt to transcend the optimist versus pessimist debate see Jenkins (1987: ch. 4) and Kiely (1997: ch. 5), and the text below.

References

Agnew, J. (1995) 'The United States and American hegemony', in P. Taylor (ed.) *Political Geography of the Twentieth Century*, Chichester: John Wiley, pp. 207–38.

Allen, J. (1995) 'Crossing borders: footloose multinationals?', in J. Allen and C. Hamnett (eds) *A Shrinking World?* Oxford: Oxford University Press, pp. 55–102.

Amsden, A. (1989) *Asia's Next Giant*, Oxford: Oxford University Press.

—— (1992) 'A theory of government intervention in late industrialization', in L. Putterman and D. Rueschmeyer (eds) *The State and Market in Development*, Boulder, Co: Lynne Rienner, pp. 53–84.

Bello, W. and Rosenfeld, S. (1992) *Dragons in Distress*, London: Penguin.

Brook, C. (1995) 'The drive to global regions?', in J. Anderson, C. Brook and A. Cochrane (eds) *A Global World?* Oxford: Oxford University Press, pp. 113–66.

Corbridge, S. (1986) *Capitalist World Development*, London: Macmillan.

Donaghu, M. and Barff, R. (1990) 'Nike just did it: international subcontracting and flexibility in athletic footwear production', *Regional Studies* 24(6): 537–52.

Economist, The (1994) 'The global economy', 1 October, pp. 3–46.

Elson, D. (1988) 'Dominance and dependency', in B. Crow and M. Thorpe (eds) *Survival and Change in the Third World*, Cambridge: Polity, pp. 264–88.

Emmanuel, A. (1982) *Appropriate or Underdeveloped Technology?* Chichester: John Wiley.

Evans, P. (1979) *Dependent Development*, Princeton: Princeton University Press.

Frank, A. G. (1972) *Lumpenbourgeoisie, Lumpendevelopment*, New York: Monthly Review Press.

—— (1981) *Crisis in the Third World*, London: Heinemann.

Freeman, A. (1991) 'The economic background and consequences of the Gulf War', in H. Bresheeth and N. Yuval-Davis (eds) *The Gulf War and the New World Order*, London: Zed, pp. 153–65.

Frobel, F., Heinrichs, J. and Kreye, O. (1980) *The New International Division of Labour*, Cambridge: Cambridge University Press.

Gereffi, G. (1983) *The Pharmaceutical Industry and Dependency in the Third World*, Princeton: Princeton University Press.

—— (1994) 'Capitalism, development and global commodity chains', in L. Sklair (ed.) *Capitalism and Development*, London: Routledge, pp. 211–31.

Gereffi, G. and Hempel, L. (1996) 'Latin America in the global economy: running faster to stay in place', *NACLA* 39(4): 18–27.

Gereffi, G. and Korzeniewicz, M. (eds) (1994) *Commodity Chains and Global Capitalism*, London: Praeger.

Girvan, N. (1976) *Corporate Imperialism*, New York: Monthly Review Press.

Glyn, A. and Sutcliffe, B. (1992) 'Global but leaderless? The new capitalist order', *The Socialist Register*, London: Merlin, pp. 76–95.

Harris, N. (1994) 'Nationalism and development', in R. Prendergast and F. Stewart (eds) *Market Forces and World Development*, London: Macmillan, pp. 1–14.

Hewitt, T., Johnson, H. and Wield, D. (eds) (1992) *Industrialization and Development*, Oxford: Oxford University Press.

Hirst, P. and Thompson, G. (1995) 'Globalization and the future of the nation-state', *Economy and Society* 24(3): 408–42

—— —— (1996) *Globalization in Question*, Cambridge: Polity.

Hopkins, T. and Wallerstein, I. (1986) 'Commodity chains in the world economy prior to 1800', *Review* 10(1): 157–70.

Hymer, S. (1982) 'The multinational corporation and the law of uneven development', in H. Alavi and T. Shanin (eds) *Sociology of Developing Societies*, London: Macmillan, pp. 128–52.

Jenkins, R. (1987) *Transnational Corporations and Uneven Development*, London: Methuen.

—— (1988) 'Transnational corporations and Third World consumption: implications of competitive strategies', *World Development* 16: 1363–70.

—— (1992) 'Industrialization and the global economy', in T. Hewitt, H. Johnson and D. Wield (eds) *Industrialization and Development*, Oxford: Oxford University Press, pp. 13–40.

—— (1993) 'The impact of foreign investment on less developed countries: cross-section analysis versus industry studies', in J. Dunning (ed.) *Transnational Corporations*, London: Routledge, pp. 111–30.

Johnson, C. (1982) *MITI and the Japanese Miracle*, Stanford: Stanford University Press.

Kiely, R. (1994) 'Development theory and industrialisation: beyond the impasse', *Journal of Contemporary Asia* 24: 133–60.

—— (1997) *Industrialization and Development: a Comparative Analysis*, London: UCL Press.

Lall, S. (1978) *The Growth of the Pharmaceutical Industry in Developing Countries*, New York: UNIDO.

—— (1993a) 'Introduction: transnational corporations and economic development', in S. Lall (ed.) *Transnational Coporations and Economic Development*, London: Routledge, pp. 1–27.

—— (1993b) 'Multinationals and technological development in host countries', in S. Lall (ed.) *Transnational Corporations and Economic Development*, London: Routledge, pp. 237–50.

Lall, S. (ed.) (1993c) *Transnational Corporations and Economic Development*, London: Routledge.

Lall, S. and Streeten, P. (1977) *Foreign Investment, Transnationals and Developing Countries*, London: Macmillan.

Lang, T. and Hines, C. (1993) *The New Protectionism* London: Earthscan.

May, H. (1975) *Multinational Corporations in Latin America*, New York: Council of the Americas.

McGuigan, J. (1992) *Cultural Populism*, London: Routledge.

Michalowski, R. and Kramer, R. (1987) 'The space between laws: the problem of corporate crime in a transnational context', *Social Problems* 34(1): 34–53.

Mitchell, T. (1992) 'The Gap: can the nation's hottest retailer really stay on top?', *Business Week*, 9 March, pp. 58–64.

Mosley, P., Harrigan, J. and Toye, J. (eds) (1991) *Aid and Power*, London: Routledge, 2 volumes.

Muller, M. (1982) *The Health of Nations*, London: Pinter.

Murray, R. (ed.) (1981) *Multinationals Beyond the Market*, Brighton: Harvester.

New Internationalist (1993) 'The new globalism: the facts', *The New Globalism*, August, pp. 18–19.

—— (1996) 'Green economics: the facts', *Seeds of Change*, April, pp. 18–19.

Ohmae, K. (1991) *The Borderless World*, London: Fontana.

Page, S. (1994) *How Developing Countries Trade*, London: Routledge.

Pearson, R. (1994) 'Gender relations, capitalism and Third World industrialisation', in L. Sklair (ed.) *Capitalism and Development*, London: Routledge, pp. 339–58.

Peet, R. (1986) 'Industrial devolution and the crisis of international capitalism', *Antipode* 18: 78–95.

Rowthorn, R. and Wells, J. (1987) *Deindustrialization and Foreign Trade*, Cambridge: Cambridge University Press.

Schiller, H. (1979) 'Transnational media and national development', in K. Nordenstreng and H. Schiller (eds) *National Sovereignty and International Communication*, New Jersey: Ablex, pp. 17–25.

Singh, A. (1994) 'How did East Asia grow so fast?', *UNCTAD Discussion Paper*, no. 97, Geneva: UNCTAD.

Sklair, L. (ed.) (1994) *Capitalism and Development*, London: Routledge.

Smith, P. (1992) 'Industrialization and environment', in T. Hewitt, H. Johnson and D. Wield (eds) *Industialization and Development*, Oxford: Oxford University Press. pp. 277–302.

Standing, G. (1989) 'Global feminization through flexible labour', *World Development* 17(7):1077–96.

Tomlinson, J. (1991) *Cultural Imperialism*, London: Pinter.

Tunstall, J. (1977) *The Media Are American*, London: Constable.

UNCTAD (1994) *World Investment Report*, New York: United Nations.

—— (1995) *World Investment Report – Preliminary*, New York: United Nations.

Vaitsos, C. (1976) *Employment Problems and Transnational Enterprises in Developing Countries*, Geneva: ILO.

Vernon, R. (1973) *Sovereignty at Bay*, London: Penguin.

Wade, R. (1990) *Governing the Market*, Princeton: Princeton University Press.

Wallace, P. (1996) 'Mixed blessings in the two way flows of foreign investment', *Independent*, 22 February, p. 24.

Walton, J. and Seddon, D. (1994) *Free Markets and Food Riots*, Oxford: Blackwell.

Warren, B. (1973) 'Imperialism and capitalist industrialisation', *New Left Review* 81: 9–44.

Williams, K., Haslam, C., Williams, C., Sukhdev, J., Johal, J., Adcroft, A. and Willis, R. (1995) 'The crisis of cost recovery and the waste of the industrialised nations', *Competition and Change* 1(1).

World Bank (1992) *Governance and Development*, Washington: World Bank.

World Trade Organisation (1995) *Regionalism and the World Trading System*, Geneva: World Trade Organisation.

chapter 3

Migration and the refugee experience[1]

Phil Marfleet

chapter 3

The notion of a 'globalised' world is often associated with the idea of flows – movements of capital, data, ideas and people. Such movements are said to produce increased integration in a world which is becoming 'a single place'. But in their enthusiasm to identify such changes theorists of globalisation invariably underemphasise or ignore efforts to control certain flows, in particular the attempts to stem or even reverse flows of people. Some Western states and transnational bodies are expending enormous energy to contain such movements, especially emigration from Africa, Asia and Latin America. By insisting that migrants from these continents remain at the margins of the world system, they seem to confirm the old binary oppositions of world affairs – those between First and Third Worlds, North and South. For global theory such divisions have been rendered meaningless: for tens of millions of migrants they are a stark reality.

This chapter examines some of the ways in which dominant states are reasserting old physical and cultural boundaries and establishing new frontiers. It focuses on the refugee, arguing that processes widely viewed as integrative of the world system are in fact productive of disintegration, stimulating mass movement of unwilling migrants. Development strategies directed by dominant states, and celebrated in 'global' accounts of world affairs, have a negative impact in regions where economic and political structures are fragile. Local states often intensify repression to protect themselves against the consequences of such strategies, especially the immiseration produced by neo-liberal economic policies. Large numbers of people may flee, and if the local state falls into crisis or collapses there may be mass exodus. Those who seek sanctuary in the West often discover new barriers to migration and aggressive campaigns directed against immigrants in general and refugees in particular.

As 'Europe' and 'the West' are redefined within the 'global' order they are increasingly identified against new (or at least reconstructed) imagined enemies. For European governments the migrant has become a focus of official hostility, often depicted as a threat to the continent's cultural integrity. As European integration advances, there is increasing emphasis on what Martiniello calls an 'ideology of cultural resistance of a besieged Europe against the South' (1994: 39). 'Fortress Europe' has been closed to all but a trickle of migrants, expressing antagonism towards the Third World of such force as to question all notions of an effective global coherence.

Global theory

In some recent studies of forced migration the refugee has been placed unproblematically in a 'globalised' context. Santel, for example (1993: 81), notes 'the process of globalisation of refugee movements' and Suhrke (1997: 217) comments on 'the globalization of the refugee phenomenon'.[2] But theories of globalisation in general do not encompass the idea of mass exclusion which characterises contemporary refugee movements. Rather, they depict an interconnected world and a coherence of political, social and cultural forms.

The level of world integration is most advanced in the perspectives of economists and of organisation and management theorists who identify the world system with the market and view the global as confirmation of the power of 'liberalising' economic forces. On this view, the world system is dynamic: market forces dissolve political difference and allow capital – mobile and adventurous – to serve populations which are 'invited in' to a new order. As part of this process, it is said, social and cultural difference worldwide is greatly reduced. Most important, state structures are dissolved, producing a 'borderless world' (Ohmae 1995).[3]

These notions have strongly influenced even theorists identified with radical critiques of world order. Harris (1995: 16), for example, describes 'the arrival of an integrated world economy' which has profound implications for socio-political structures. Both capital and labour now circulate with increasing freedom, he argues, restrained only by the enfeebled state structures of an earlier era. Lauding these developments, Harris (1995: 228) concludes that from within the new order, 'world interest and a universal morality . . . are struggling to be reborn'.

Some social and cultural theorists have adopted a more restrained approach, identifying complexities and even contradictions within globalism. Theorists such as Giddens (1991), Harvey (1989) and Mann (1992) identify compressions of space and time, and the new 'social spaces', created by globalising processes. They examine expressions of difference which may emerge as such processes take hold, emphasising the possibility of diversity within unicity. So too with Featherstone's interest in local/global (what he dubs 'glocal') relationships (Featherstone 1992).

Even these approaches are expressed within the global paradigm, however. They speak of flows, movements across old boundaries and of the social flux within which new identities are forged – but remain largely silent on mass movements of people. When Arrighi (1996), for example, writes of the abstractions of increased mobility and the redefinition of space – 'a space of flows, rather than a space of places' – his reference is solely to commerce, to financial transactions and to capital movement. Migration is at best a passing reference and forced migration is invariably ignored: Robertson's *Globalization: Social Theory and Global Culture*, which has become a defining globalist text, entirely omits both issues.

Most theorists of globalisation also make the Western experience their point of reference. Almost the whole literature on globalisation views the new order as an outgrowth of European traditions of economic, social and political advance in which the long history of interactions with the Third World does not require serious examination.[4] Absent are references to the contestations of Western dominance which have been enormously important in shaping the contemporary world, including European society. Today's mass movements of people within and from the Third World, which in themselves raise critical questions about dominant world models, hardly enter the globalist picture.

Mass population movements do demand a world frame of reference. They are as much a part of a notionally more fluid world as flows of capital or information. They demonstrate, however, that there are serious limits to the idea of an integrating global system, for they reveal starkly the ways in which new obstacles to free movement are being erected, especially *vis-à-vis* the people of the South. These contradictory aspects of change at a world level largely go unnoticed within the enthusiastic noises about world integration generated by globetalk.

The refugee

In the late 1950s, says the United Nations High Commissioner for Refugees (UNHCR), there were some 2 million refugees; by 1995, the organisation estimated that numbers had reached 27 million (UNHCR 1995). Other sources quote far higher numbers: according to Harris, there may be as many as 70 million refugees, with possibly the same number in flight within countries (Harris 1995: 120). To these figures may be added the numbers judged to be 'at risk', among whom many are potential refugees. According to a US government study, in 1996 some 42 million people were in extreme physical danger, mainly as the result of regional political conflicts (*Guardian International* 14 April 1996).

These figures provide only the most crude measures: not only is the process of counting often haphazard but the notion of 'refugee' is differently constructed among states and transnational bodies. Headline statistics none the less give a sense of the unprecedented scale of migration and of the vast numbers of those whose insecurity leads to flight.

Who should be counted? The matter of definition is a battleground: as Tuitt (1996: 2) has recently pointed out, states and international bodies expend much energy in setting refugee status within a strictly defined legal context, 'seeking to portray the phenomenon of refugeehood as reducible to a legal definition'. Much rests upon their interpretations of the 1951 Geneva Convention Relating to the Status of Refugees. This document and its 1967 Protocol lay down a definition to be applied specifically to individuals which turns upon subjective interpretation of the asylum-seeker's experience of persecution. There is no conception of the collective refugee, notwithstanding that the Convention introduces a notion of persecution which implies oppression of whole groups on the basis of their 'race, religion, nationality, membership of a social group or political opinion'. Collinson comments on the utility of this approach for the Western governments which supervised drafting of the Convention in the early 1950s. Then, with refugees associated with the European dimension of Cold War relations, there was 'a certain political advantage for Western receiving states when it could be accorded to refugees from the Eastern bloc' (Collinson 1994: 20). For Suhrke (1997: 219), the focus upon European concerns, together with disinterest in contemporary forced migration in the Third World, amounted to a 'Eurocentric orientation' of the early refugee regime.

Although there is now formal recognition that 'refugee-producing' countries are overwhelmingly in the Third World, the Convention has not been modified. Wider definitions, including those adopted by the Organisation of African Unity, incorporate the idea that 'every person' threatened by a range of external or internal threats should be offered asylum (Tuitt 1996: 12). But laws based upon the Convention continue to exclude assistance for all involuntarily displaced persons and insist on the principle of 'alienage' – that the 'legitimate' refugee must, in the words of the Geneva Convention, be 'outside the country of his [sic] nationality'. Displaced persons who have not crossed national borders cannot be granted refugee status. The overall effect of these legal principles, Hathaway concludes, is that need is defined 'in terms which exclude most refugees from the less developed world' (Collinson 1994: 21).

As against these narrow definitions, the notion of 'forced migration', of those coerced into flight, is the only approach which encompasses the predicament of most asylum-seekers. Such an idea has long been resisted by Western governments keen to differentiate refugees from other categories of migrants. During the 1950s and 1960s, migrants were usually identified on the basis of crude categories: 'economic' migrants, those seeking family reunion, recognised refugees, asylum-seekers and 'illegals'. Today migrants are less exceptions to a rule in which stable settled communities are rooted in 'place of origin' than part of the process in which capital, information and ideas do move more freely across national boundaries and in which large numbers of people are both induced and coerced to migrate. Differentiation between 'economic' migrants, 'refugees' and others becomes meaningful only for those most determined to perpetuate systems of exclusion.

How should we understand the refugee in this context? Traditional notions of 'pull' and 'push' factors in migration may still be useful. It has often been argued that mass migrations are a result of movement into expanding economies or a function of expectations that migration can provide enhanced 'life chances'; this constitutes a 'pull' factor. On the other hand, intolerable conditions in places of origin similarly make for a move, even when conditions elsewhere might be less attractive than hitherto – the 'push' factor. When this approach takes account of political and social as well as economic factors it has an explanatory value: refugees are those among whom the 'push' factor is absolutely decisive. The refugee is a woman or man with the narrowest range of choice, usually because specific local conditions have made for exclusion. Such conditions may be explicitly 'political' – relating to repression of particular parties, organisations or individuals, ethnic, 'racial' or religious groups, or to people of a particular sexual orientation. Equally, causal factors may be primarily economic – related to immiseration, landlessness, famine or environmental collapse. Even the US Department of Labor – not usually identified with refugee causes – comments that 'increasingly, both pure refugees and purely economic migrants are ideal constructs rarely found in real life; many among those who routinely meet the refugee definition are clearly fleeing both political oppression and economic dislocation' (Papademitriou 1993: 212–13).

Uneven development

Hathaway comments that 'contemporary international refugee law is marginal to the protection of most persons coerced to migrate' (Collinson 1994: 21). The vast majority of such people are to be found in Africa, Asia or Latin America, or in flight from countries of the South to those of the North. In 1992, 90 per cent of refugees and asylum-seekers officially recorded by the US Committee for Refugees were located in the Third World (*ibid.*: 19). US Intelligence has estimated that tens of millions are additionally at risk in Africa, Central Asia and the Middle East: in Europe, only Bosnia has been viewed as a zone which might produce similar humanitarian crises (*Guardian International* 14 April 1996).

The UNHCR believes that this uneven pattern – of Third World crisis and flight towards the West – is likely to become more pronounced:

> The world's refugee burden is carried by the poorest nations: the 20 nations with the highest ratio of refugees have annual per capita income of $700 . . . The refugee situation in the South is already dramatic, and its tragic dimensions are bound to spread in the decades to come.
>
> (Widgren 1993: 89)

This development is linked intimately with changes in the world economy which challenge the theory of a globalising world. Globalisation theory depicts a world integrated by capital flows. Movements of finance and investment are said to 'seed' growth across the continents: 'virtual' banking speeds money through a new network of financial centres, while multinational companies (MNCs) invest far from their country of origin, drawing formerly 'undeveloped' regions into the world's industrial networks. Uncontrollable market forces drive this process: the globalisation literature is replete with references to the 'juggernaut' or 'Leviathan' of integration. Hirst and Thompson (1996: 195) sum up the picture as one of 'a world internationalized in its basic dynamics'.

A series of analyses has recently questioned this globalist orthodoxy.[5] In the present context two issues are of special importance. First, globalisation sees capital as flowing unproblematically across a world market. But capital is not singular: flows of money through banks, investment houses and capital markets are not the same as flows of commodity capital or the productive capital associated with long-term investment, especially in industry. Although the volume of money exchanged transnationally has increased enormously, and there has been a revolution in means and speed of financial transactions, no such situation obtains in other areas of the world economy. As Hoogvelt (1997) shows and Kiely demonstrates elsewhere in this book (Chapter 2), investment in infrastructure and industry in Africa, Latin America and most of Asia has in fact *decreased* dramatically over the past thirty years. The Third World may be integrated into a wider system by those who operate on 'emerging' capital markets or who speculate against local currencies, but it receives less of productive investment than even in the colonial period.[6] These regions are exposed to flows of finance, fluctuating commodity prices and volatile currencies which often destabilise local economies already weakened by *loss* of long-term investment.

A second problem is that of the ahistorical character of globalisation theory. Globalisation largely ignores the processes by which capitalism became general; most important, it fails to address the means by which the nation-state became a means of assuring Western hegemony. In fact the structures of power devised by colonial administrators and established across Africa, Asia and Latin America produced highly centralised but inherently unstable structures. Almost invariably, social and cultural differences were exploited to the advantage of colonial powers and later of specific local ruling groups. Under pressure, these structures show an extreme fragility. Weakened by declining long-term investment, they are especially vulnerable to the capital flows associated with 'globalisation' of finance, becoming sites of intense conflict around which many of the most severe crises of forced migration have developed.

For Overbeek (1993: 27) the key global factor precipitating refugee crisis is the 'internationalisation of the state in the context of the transnationalisation of the political economy'. As part of the 'liberalisation' process endorsed by most global accounts, transnational economic agencies such as the International Monetary Fund (IMF) impose reform policies on Third World countries. Attempts to reduce radically state involvement in the economy invariably undermine already enfeebled structures of authority and may precipitate generalised social crisis. The limited authority of the state is undermined and sections of the state apparatus disintegrate, leading to increased repression, factional conflict and sometimes to huge population movements. What Zolberg *et al*. (1989: 44) call 'the exit from the state' is often a movement into exile.

In some regions of the Third World, dislocation of local economies and political structures is so general that transnational bodies have largely abandoned systematic development policy. Cox (1995: 41) comments that here, 'Policies to promote economic development have been very largely displaced in favour of what can be called global poor relief and riot control.' In such marginalised regions, sudden crises of food availability or the eruption of social or political conflicts can have a traumatic impact. Millions of people feel compelled to flee the zone of crisis, producing what Cohen and Joly (1989: 6) have described as 'mass exodus from the wastelands'. External intervention, especially by powers attempting 'stabilisation', may have the effect of greatly intensifying such movements.

Worldwide, the majority of refugees emerge from such areas and seek asylum in nearby countries which are often equally enfeebled. Whole regions can be affected by forced population movement: the Horn of Africa, Central Africa, Central America, Central Asia. Flight generates flight as fragile networks of support are weakened further by refugee influxes. Refugees can appear to overwhelm such regions, leading organisations such as the UNHCR to describe them as 'awash' with the displaced (*Guardian* 15 May 1995).[7]

To the extent that they are an index of change in the world economy, the coerced movements of millions of people of the Third World do describe aspects of a 'global' process – but this is not the process described by theorists of globalisation. Rather, it is one consistent with a further exaggeration of the processes of combined and uneven development characteristic of capitalism since its first phase of expansion from Europe. The refugee phenomenon therefore does not speak of world 'order' induced by dissolution of political structures but indicates a systemic *dis*order rooted in the uneven and partial character of change. The functional 'fit' implied by organic models of interlocking markets and transnational networks is less appropriate than the notion of processes which root new contradictions and create new sites for conflict.

'Waiting for a chance'

Numbers of refugees in Third World countries are vastly greater than those which have caused refugee 'panics' in Europe and North America in the 1990s. In Guinea, for example, by the early 1990s, one in every sixteen members of the population was a recognised refugee; in Malawi the figure was over one in ten (Collinson 1994: 19). By the late 1980s, the number of Afghan refugees in northern Pakistan had reached some 3 million; about the same number had fled to eastern Iran. Despite the local impact of these and similar migrations they have been relatively unhindered by 'host' states and societies.

Refugee groups such as Palestinians and the Vietnamese 'boat people' have sometimes faced extreme hostility within their regions of origin but many Third World countries have accepted recent refugee movements without widespread opposition and – with some important exceptions – largely without hostile state intervention. Relatively free movement may suggest an ineffectiveness of the local state in sealing national borders. More often it indicates the arbitrary nature of frontiers drawn by colonial powers across ethnic, religious or linguistic unities, where the refugee presence is a confirmation of local identities – as in the case of Somali refugees in Kenya or Ethiopia, or of some sections of Kurdish or Afghan society. In general, however, there seems to be a preparedness of the mass of indigenous society to accept displaced people.[8]

Refugees are none the less seldom absorbed into the mainstream of the host society, whether in exile or 'displaced' within their country of origin. Most refugees in Third World countries survive as marginals: tolerated but on the edge of societies in which the mass of the population enjoys only a precarious existence. There is often pressure to move on: to advance the project of 'Return' or to find a more desirable place of exile.

Here patterns of change at the world level have had a profound impact. One aspect of change often identified with globalisation is in means of transport and communication. Although the mass of the displaced may be scattered, or may remain for years in isolated camps and holding centres, many now have freer access to major urban centres. Here they are able to join exile communities with increasingly sophisticated communications networks. This is especially true in continental or regional centres which are themselves elements in an increasingly interconnected world. Cities such as Cairo, Nairobi, Istanbul, Peshawar, Mexico City, Rio de Janeiro, Bombay and Hong Kong have become centres of refugee activity in which exile groups attempt to establish a framework for social and political activity which is often directed to onward movement. Since the collapse of the Soviet Union and the Second World bloc, Moscow has become a further such centre, with an estimated 200,000 asylum-seekers from the Third World waiting to move to Western Europe and the United States (*Guardian* 29 December 1995).

In such cities information about the international legal regime and specific country practices may be more freely obtained. One result is that established exile communities increasingly share certain features, especially an orientation on human rights issues and on the organisations associated with them. In a world in which political movements, governments and non-governmental organisations (NGOs) routinely invoke human rights principles, the notions of a defined refugee status and an associated package of rights have become part of the exile's worldview. For those who move out of areas of crisis into the insecure but semi-permanent world of exile, 'rights' are an important focus – an abstraction that gives refugees a shared identity and often some sense of a common project, part of what Suhrke (1997: 218) calls 'globalization of the politics of refugee policy'.

The demand for rights may be focused on government bodies or on NGOs: the UNHCR, for example, is often a target of intense lobbying by exile groups on the expectation that it can effectively mobilise on their behalf. As Cooper (1992) has shown in a fascinating study of refugee communities in Cairo, a lot of energy may be expended on efforts to negotiate the networks of diplomats, counsellors and administrators who make up the human rights 'community' to be found in regional centres. For the Ethiopian and Eritrean refugees studied by Cooper, community life is focused on 'waiting for a chance' of further migration. A fund of information is gathered relating to refugee law, documentation, specific country practices and means of approaching the asylum 'gatekeepers' at consulates and NGO offices. A true vocabulary of the refugee condition – a key part of the discourse of exile framed within refugee communities – has come into existence, everywhere translated into specifics which can comprehend the intensely felt needs of each group.

It is in these contexts that established communities of exile have developed what Axford (1995: 27) describes as 'a sharpened awareness or consciousness of global pressures'. Here the perception that 'world society' has a place for the refugee can help to engender the basis of what Axford (1995: 27) calls a 'cognitive global order'. This is not to suggest that all or even many refugees apprehend a global system but that the experience of migration and of exclusion can induce growing consciousness of world structures much beyond the national frontier. Paradoxically, it may be a continuing orientation on Return – to a specific national 'home' – which helps to shape such perceptions.

The focus on universal principles does not merely induce the asylum-seeker to identify as an abstract 'refugee'. One effect of interaction with the human rights network is to bring an awareness of the constructed nature of legal approaches to asylum, especially of their inconsistency and divisiveness. Cooper (1992: 49) comments: 'the system itself generally appears to members of the refugee community as arbitrary, inconsistent, occasionally characterised by favouritism, and impossible to comprehend fully . . . '.

Refugee communities become highly sensitive to means by which governments and NGOs define an appropriate claim for asylum. Subtly different interpretations of international law, the varying influence of NGOs at different times and in different countries, and the personalities of officials, all play a role in shaping asylum-seekers' perceptions of how to represent themselves. Cooper (1992: 48) observes that whatever the particular circumstances of their own exile, refugees may feel compelled to offer a 'carefully considered and constructed life history'. When perceptions of the required history prove to be inaccurate, asylum-seekers may be 'trapped . . . in a form of legal and social limbo' (*ibid.*: 50). Here the attempt to invoke universal principles leads only to a more traumatic experience of exile.

Human rights networks mobilise around universal principles but often operate under conditions which divide and isolate refugee communities. Exile groups may seek special relations with governments and NGOs which they use to privilege community members and even mobilise in disputes with other groups. Where very large numbers of refugees congregate, conflict around such issues may become intense. In some east African cities, for example, where refugees from Ethiopia, Eritrea, Somalia, Sudan, Uganda and various central African countries have gathered, competition for access to information about the latest 'chance' for further movement is said to be particularly intense.[9] This situation may be made more complicated when governments attempt to use refugee communities, inter-communal and intra-communal frictions, in foreign policy disputes. Under such circumstances the experience of exile becomes particularly contradictory, for the shared status of 'refugee' itself generates an acute sense of difference.

The notion of onward movement towards desired places of asylum may orient refugees outwards; equally, aspects of 'global' change may focus concerns more intently on the place of the initial exclusion. From Third World cities it is increasingly easy to communicate with compatriots elsewhere through telephone, fax, Email and the Internet: Suhrke (1997: 218) comments on their 'revolutionary impact' on international solidarity networks and refugee groups. This can generate much faster recuperation of the social, cultural and political communities which in the past have often been the first casualties of exile. This 'global' change can facilitate a particular localism: the orientation on Return which is common to most refugees during the early phases of exile may be greatly focused by such systematic contact with the physical place of exclusion.[10]

The mega-cities of the Third World have become way-stations. Here large communities of refugees find themselves among shifting populations of local rural migrants, business people, expatriates and tourists. For some theorists of globalisation such locations constitute a cultural 'hyper-space' – a milieu in which localism loses its frame of reference as the means of generating authenticity and more elaborate identities are shaped and reshaped (Jameson 1991). But refugees, coerced into migration, seldom find such spaces as congenial as those employed by governments, transnational companies and NGOs like those which administer disaster relief and monitor refugee movements. These 'global travellers' invariably return to a home base. For most refugees, especially those in the early phases of exile, return or onward movement remains a compelling aim. The spaces they occupy are not free zones of movement. They are new points of departure, sometimes permanent barriers to further movement, almost never 'home'.

Asylum

Greater ease of communication is often seen to enhance world integration. This is largely an illusion, for social structures at world level demonstrate extreme rigidity. Even modification of political boundaries takes place on the basis of old patterns of exclusion. The construction of 'Fortress Europe' demonstrates how readily traditional structures can be reasserted.

Pressures to migrate from the Third World, observes King (1995: 26), have produced a particular response in the desired countries of destination: 'The reaction of the West is usually to pull down the shutters and deny people who have lost their original place in the world the chance to find another'. He continues:

> The rules of modern migration are clear. 'Desirable' migrants with skills, education and capital are let in; 'undesirables', illiterates, poor people from different cultures, religions and 'races' are filtered out. Globalisation is a process of social exclusion.
>
> (King 1995: 26–7)

From the mid-1980s, numbers of asylum-seekers arriving in Europe increased four times to some 300,000 annually (Collinson 1994: 19). Pressures to migrate were becoming irresistible; at the same time, perceptions of Europe as a place of asylum had been heightened, especially as it became more widely known that Vietnamese and Chilean refugees had earlier been accepted by a number of European countries (Santel 1993: 81).[11] Some migration routes had been well established by those earlier deemed 'economic' migrants but increasingly refugees progressed through the networks being established in Third World cities where onward-moving migrants were familiarised with opportunities, or perceived opportunities, in Europe.

In addition, an increasing number of refugees moved directly from country of exile to the desired country of asylum, a move facilitated by changes in transportation that placed most of Europe within a few hours' flying time. This was a radical change: until the 1980s, travel from many formerly distant regions of the Third World had been complex and usually required refugees to move through a limited number of Third World centres. The expansion of transport systems which accompanied economic change and the massive extension of tourism opened new possibilities. Much of the Middle East, South Asia and the Horn of Africa – the main 'refugee-producing' regions – could now be reached by regular scheduled air flights, making every European country potentially a country of first asylum. This is the sense in which Hocke has described refugees of the 1980s and 1990s as 'jet people' who have succeeded the 'boat people' of the 1970s (Santel 1993: 81).[12]

The majority of refugees from the Third World who gain access to the West have concentrated in Europe's metropolitan centres. Here, some are absorbed into established communities of compatriots who followed earlier migration paths. Others have moved into more marginalised layers of impermanent, shifting populations. In this respect, London, Paris or Berlin have much in common with the Third World megacities. Communication with both country of origin and with refugee communities in other such centres is often more systematic than with the wider European society. Changes in means of communication allow interaction between individuals who are in fact formally linked by their *distancing* from the country of exile and from other physically distinct communities.[13] In some cases relationships across refugee diasporas may be intensified by access to media which broadcast from the region or country of exile, bringing together large numbers of refugees in an environment in which 'home' seems tantalisingly within reach.[14]

These changes have been accompanied, however, by greater marginalisation within countries of asylum. During the 1980s asylum-seekers attempting to enter Europe encountered societies marked more and more systematically by structures of exclusion. On arrival, those unable to navigate complex and intimidating bureaucratic procedures might be expelled. Many of those admitted temporarily were imprisoned, imposing upon the exile a further exclusion. Thus people of diverse nationalities, ethnic groups, religions and languages found themselves sharing prison cells, detention centres, interrogation rooms and immigration courts. Such places of containment – created specifically for 'aliens' – make a pointed statement about the fluidity of space celebrated by much recent social theory.

In most major European cities refugees also encountered complex transnational and national networks engaged in human rights campaigning; refugee welfare, education and community support; legal advice and advocacy. For a minority of refugee activists there might even be salaried positions as community workers, counsellors, translators and advocates. Within such networks, the vocabulary of human rights, of universal principles, again took on special significance. Double exclusion made for circumstances in which the particular predicament might be eased by reference to universal standards authored in the West and which the refugee challenged the host society to implement. Now the refugee – with all her or his specific experiences of national exclusion – was the bearer of putative 'global' values.

In Europe, however, the refugee was even more exposed than in the Third World city. In order to obtain asylum rights within a specific national state it became crucial to read the local definition of refugee status correctly. The mediatory layer of community activists, translators and counsellors became a key source of knowledge of the host society at its official level and as such was able to develop privileged relationships for certain communities, or for ethnic or political groups. One repeated grievance of successive groups of asylum-seekers in Britain, for example, is that each must build new relations with the immigration authorities. In part this has been achieved by contesting relationships established by long-standing groups which have their own social, political and other allegiances.[15] Here, the increased pace of forced migration and the reluctance of Western states to accommodate it makes for differentiation of refugee communities which in other ways are acutely conscious of their shared circumstances.

By the late 1980s the major European cities were receiving asylum-seekers from throughout the Third World. According to the UNHCR's very conservative estimate, numbers of refugees worldwide rose by 70 per cent to 17 million between 1985 and 1991 (Collinson 1994: 19). In fact, only a fraction of these numbers reached Europe: over 90 per cent remained in the Third World, together with an equal proportion of 'displaced' persons. The increase in asylum applications in Western Europe during the second half of the 1990s represented under 5 per cent of the total rise in numbers of refugees worldwide (Collinson 1994: 19). None the less, by the late 1980s, most Western European countries had experienced 'asylum crises', 'refugee panics' and vigorous new campaigns of exclusion. Within a few years Fortress Europe had been walled off: in the face of an increasingly urgent need for asylum the European Union had become 'a club for the racially privileged' (*The Observer* 7 January 1996).

Fortress Europe

Castles (1993: 19) comments that 'Migratory chains, once established, continue, even when the original policies on which they were based are changed or reversed.' Most recently, this has been expressed by the pattern of 'economic' migration among those encouraged to move from colonial and former colonial states to Western Europe during the 1940s and 1950s. When European states began to restrict movement in the 1970s, and later attempted to block entry, it proved impossible to break the migratory chain. As Weiner notes (1997: 99), European governments 'soon found that migrants were people with wills of their own'. Many had 'settled', at least temporarily, families required reunion, and there was growing pressure for asylum from those coerced by Third World states in various phases of crisis. In similar fashion, the changing policies of labour-hungry states such as Germany, which had drawn in vast numbers under its 'guest worker' programme, could not reorder the migratory pattern.

By the late 1980s even southern European countries which had been exporters of migrants, such as Italy, Spain and Portugal, were experiencing much-increased immigration from Africa and Asia. In addition, flows of Third World migrants into Western Europe had been augmented by those from Eastern Europe, where some states were at the point of collapse. These developments coincided with the most important collective project within Europe in over fifty years: the attempt at collaboration among the nation-states of Western Europe.

The European Union (EU) had its roots in the reconstructionist perspectives of post-war Europe. By the 1980s, however, it had a different agenda, reflecting the concern of its dominant states to assert their regional presence and to intervene effectively within a changing world economy.[16] Martiniello (1994: 33) comments in this context on 'the permanent primacy of the economic dimension in the construction of the European Community', the task of which was 'to complete the internal market [within Western Europe] as soon as possible and to assure the conditions of its efficiency'. By the late 1980s the aim of integration had been given added urgency by collapse of the Second World associated with the former Soviet bloc, the trumpetings of New World Order which accompanied US assertions of global hegemony, and by the emergence of the assertive capitalisms of East Asia.

Some of the complexities of interacting processes at the world level are illustrated in this conjuncture. The campaign for European unification reached new heights at the moment when migratory flows intensified. These developments were not discrete: they were intimately connected at the 'structural' level and at the level of socio-political interaction between Europe and the former colonial world.

The idea of a European unity began to take on more overtly political and cultural dimensions. As the ideologues of pan-Europe attempted to define their project they turned more and more to imaginings of a Europe defined by those whom it excluded. A European super-state was to be a 'community of destiny', 'a European spirit', above all, a Europe of peoples defined by a perceived common Christian heritage (Martiniello 1994: 38). The corollary was that peoples from societies associated with other traditions, especially other religions, were not of Europe and were not to be of Europe. Such notions drew on long-established racist ideologies with their roots in the earliest period of colonial expansion, in the emergence of industrial capitalism and the consolidation of the European nation-state.

For several centuries, Europe has maintained relations with a Third World which it both marginalised and, at the cultural level, sought to incorporate. This produced what Schwarz (1996: 5) identifies as a set of 'powerful exclusions', which in the case of England 'were connected by an intimate set of relations to the workings of colonial rule'. In their attempt to situate Europe favourably in a 'globalised' world the ideologues of the EU have intensified these exclusions. Prompted by the search for a new ideology of European coherence, they have easily accommodated to pressures exerted by populist racisms of the extreme right. Delanty (1995: 150) points to the importance in this context of neo-fascist attempts to organise around imagined threats from the Third World, especially 'the xenophobic spectre of not only a Muslim-dominated world but an "Islamisation of Europe"'. In this sense, the idea of a New Europe draws upon the same fund of imaginings that earlier provided material for National Socialist ideologues of an homogenous European culture.

Collinson comments (1996: 40) that by the early 1990s instability on Europe's eastern and southern borders, and an associated increase in the numbers of those seeking asylum in EU states, had fuelled 'a growing paranoia complex in western Europe which centred on apocalyptic images of a Europe under siege'. Leading politicians started to construct what the *Guardian* described in 1991 as 'doomsday metaphors' which represented Europe as the likely victim of growing external threat (*Guardian* 16 November 1991). Of these threats the 'demographic time bomb' of North Africa was judged most serious. Within this region, it was argued, youthful populations looked outwards to Europe to satisfy their economic ambitions: 'These demographic "gradients" conjure a picture of a European fat city on a hill, up whose slopes will rush hundreds of thousands of the desperate, the angry, and the enterprising' (*Guardian* 16 November 1991).

Just such attitudes underlay discussions at a series of secretive meetings which had already been held by EU foreign ministers and Eurocrats to develop collective policy on immigration in general and on refugees in particular. In a prescient article written in 1989 Cohen and Joly commented on the contrast between intra-European disputes and the consensus which emerged at these discussions. Formally, they noted, European governments differed widely in political complexion and found it intensely difficult to agree on the bases for economic and political collaboration: 'they bicker endlessly about lamb imports, agricultural subsidies, monetary policy and the potential shape of a possible political union' (Cohen and Joly 1989: 15). Yet a uniform policy on Third World asylum-seekers had quietly emerged, shaped largely outside public forums. Cohen and Joly concluded:

> It has emerged because of the broad similarity of political conditions facing European governments – the growth of a populist right wing inside each country and the emergence of a global crisis in the generation of a third world refugee population.
>
> (Cohen and Joly 1989)

Of the various EU agreements on refugees the Schengen Treaty was the most far-reaching (Santel 1993: 3). Signed by France, Germany and the Benelux countries in 1985, it was later joined by a series of other states. The agreement aims at passport-free movement within the territories of signatory states – so-called 'Schengenland'. It also introduces strict controls on immigration, including strengthened external borders and a massive central data base for recording attempts at illegal entry. EU officials and ministers have established a liaison group to develop draconian legislation in support of the agreement. In 1992 this approved arrangements to expel asylum-seekers who had passed through a 'safe' third country. In 1995 it formulated proposals to allow EU states to expel even those officially recognised and defined as 'refugees' under the terms of the Geneva Convention.

Thus as refugees have more and more urgently invoked the universal principles that were once inscribed in international law by European states, the latter's political leaders have hastened to reconstruct them. Tuitt has described these erosions of legal identity as marking 'the death of the refugee' in Europe (Tuitt 1996: 1, 120).

The attempt to construct a *cordon sanitaire* around the EU is, however, less a response to requests for asylum than a key aspect of efforts to make a new pan-European identity. As the EU takes on a more political dimension its leaders and associated academics have searched for means of defining a socio-cultural meaning for the bloc. At the moment of its emergence onto the world scene as a collective presence, pan-Europe is being defined by the narrowest measures of *national* identity, with an imagined European culture defined in relation to the continent's historic Other, identified above all in the Third World refugee.

Smith has emphasised the difficulty of uncovering the elements of 'Europeanness':

> There is no European analogue to Bastille or Armistice Day, no European ceremony for the fallen in battle, no European shrine of kings or saints. When it comes to the ritual and ceremony of collective identification, there is no European equivalent of national or religious community.
>
> (Smith 1992: 73–4)

Smith (1992: 74) asks whether it is possible for the new Europe to arise without 'myth' and 'memory'. The ideologues of pan-Europe have attempted just such imaginings. In a Europeanised version of Samuel Huntington's 'clash of civilizations', EU ministers have argued that the continent's main threat no longer comes from the communist East but from behind 'a new fault line' which has allegedly replaced the Iron Curtain.[17] This, it is said, runs along the Mediterranean, dividing southern Europe from North Africa and from the menace of Islamic societies. In February 1995, Spanish Foreign Minister Javier Solana insisted that economic crisis in North Africa, together with population growth, provided 'all the ingredients for the conflict between Islam and Europe that has made up so much of the unhappy history of the Mediterranean' (*Independent* 8 February 1995). What allegedly gave the prospect of conflict a special edge was an existing North African migrant presence in Europe that could not be allowed to consolidate. Here, the Other Within is a bridgehead for perceived threats to European coherence and stability. Islamic society in general – in the Middle East and within Europe – was becoming the focal point within this racist discourse. Balibar has observed:

> The two *humanities* which have been culturally and socially separated by capitalist development – opposites figuring in racist ideology as 'sub-men' and 'super-men', 'underdeveloped' and 'overdeveloped' – do not remain external to one another, kept apart at long distances and related 'only at the margins'. On the contrary, they interpenetrate more and more within the same space of communications, representations and life. Exclusion takes the form of *internal exclusion at world level*: precisely the configuration which, since the beginnings of the modern era, has fuelled not only xenophobia or fear of foreigners, but also racism as fear and hatred of *neighbours* who are near and different at the same time.
>
> (Balibar 1991: 15, author's emphases)

In November 1995, the EU concluded the Barcelona Declaration, known as Club Med – the first agreement with North African states to provide long-term European funding for infrastructural development. The quid pro quo was an undertaking by Arab governments to control emigration. Heralded by the European press as the start of a new era of dialogue between 'ancient rivals', the deal was in effect an attempt to seal off the EU to migrants from the Arab states (*Guardian* 25 November 1995). Such developments seem to confirm Delanty's observation that 'Europe is becoming a fortress with the Straits of Gibraltar and the Bosporus as moats and the Third World being held at arm's length.' This is no less than a European account of Huntington's global 'clash'.

The Barcelona Declaration is one important expression of the attempt to identify the EU as an *essential* Europe, one defined against a cultural bloc seen precisely as an aspect of European memories – those of 'ancient rivalry'. It is appropriate that Smith (1992: 76) answers his rhetorical inquiry by concluding that the European project's quest for collective identity is likely to end in the assertion of 'reactive' identity against the Third World on the basis of 'cultural and racial exclusion'.

Human cargo

Sealing of borders to many asylum-seekers has greatly intensified exploitation of the most vulnerable, with a huge increase in attempts to smuggle refugees, described in a recent US government report as 'a growing trade in human cargo . . . made possible by staggering levels of official corruption' (*Guardian* 29 December 1995). One outcome has been a series of tragedies in which asylum-seekers attempting to enter Western Europe hidden in trucks and containers have suffocated or starved to death, while hundreds have drowned when attempts to land on Europe's Mediterranean coastlines have ended in catastrophe.[18] The new measures of exclusion have also produced the syndrome of the 'refugee in orbit', the 'pinball' or 'serial' refugee who is a victim of repeated refusals and expulsions.[19]

The EU now reaches beyond its borders in efforts to prevent asylum-seekers reaching points of entry. Central European states, for example, have been urged to form a 'buffer against the hordes from the East' (Overbeek 1993: 32).[20] European officials have intervened directly in many Third World countries: Dutch immigration officers, for example, have been stationed at Nairobi Airport to screen potential Somali refugees and return them to UNHCR camps in Northern Kenya (*ibid.*: 32). This is consistent with attempts by many Western states to intervene in the refugee-producing regions, both by policing movements of asylum-seekers directly and by using the UNHCR and other NGOs as regulatory bodies. According to Weiner (1997: 100) this practice is increasingly successful.

Here the NGOs' role as 'gatekeepers' within the rights network is increasingly that of keeping potential openings to the West firmly shut. Cooper's study of refugees in Cairo shows that many asylum-seekers view the UNHCR itself – the leading transnational advocacy organisation for refugees – as essentially hostile to their cause. This picture is confirmed by informal evidence from many refugees from East and central Africa where Western-funded NGOs have been particularly active in recent years.[21]

But the most vigorous assault on the refugee is within European society. Here the British state has distinguished itself by initiating a campaign of particular hostility. It has inculcated a 'culture of disbelief' among immigration officials, as a result of which the vast majority of applications for asylum have been rejected. In 1990, 23 per cent of all applicants for asylum in Britain were granted refugee status and 60 per cent were given 'exceptional leave to remain'. By 1995 the figures had shrunk to 4 per cent and 18 per cent respectively and the British police were carrying out large numbers of deportations (*Guardian* 2 January 1996). At the same time, the government portrayed refugees as illegal aliens parasitic on the British benefits system and attempted to deny them access to state support. Even the mild-mannered UNHCR has been moved to oppose the policy as a violation of international treaty obligations.

Under its Asylum and Immigration Act, the British government has provided for criminal charges to be brought against employers alleged to have hired immigrants who do not have a regularised status (*Guardian* 13 March 1995).[22] In France a similar attempt has been made to initiate 'popular regulation' of immigrants. A bill introduced in February 1997 aimed to compel French citizens to inform police of the presence of foreign 'guests', threatening penalties for non-compliance.

Local and global

Attempts to induce a culture of hostility towards asylum-seekers are consistent with the creation of a series of 'official' renewals of racial exclusivism which are taking place across Europe. Here the role of the state itself is significant. Balibar has argued:

> [M]odern racism is never simply a '*relationship to the Other*' based upon a perversion of cultural or sociological difference; it is a relationship to the Other *mediated by the intervention of the state* . . . it is the state qua nation-state which actually produces national or pseudo-national minorities [which] only exist in actuality from the moment when they are codified and controlled.
>
> (Balibar 1991: 5, author's emphases)

European states have gone to great lengths to reassert understandings of the character of 'non-Europe' as part of their means of relocating nation and the European collective within what they see as a new globalising environment. It is highly significant that the minority which they have sought to 'codify and control' most vigorously is that of the refugee. Conscious of new pressures exerted by the 'world of flows', including pressures on constructions of national identity, they have felt it necessary to invigorate the racialisation of their cultures. Significant in this respect are European governments' efforts to induce the wider (white) society to police asylum-seekers, intensifying consciousness of an imagined threat which is both external and internal.

As the project for a collaborative Europe has become more pressing the EU itself has turned to national traditions of exclusion to find the vocabulary with which to define a continental identity. The form produced by this interaction with 'global' forces is a localism which makes of pan-Europe no more than recycled myths and imaginings of specific nationalisms. As Miles has observed: 'the boundary of "our" economic and political field has been extended, necessitating an extension of the boundary of the "imagined community" beyond that of each nation state (and hence the *renewal and reconstruction of the idea* of Europe)'. (Miles 1993: 51, author's emphases).

In this reconstruction it is the asylum-seeker as one important expression of a changed world – a more interconnected but also more volatile and insecure environment – who bears the main weight of definition. The refugee is made to carry the history of earlier European racisms but is also now charged with responsibility for the perceived instability and disorder of a crisis-wracked world. In Europe and in other dominant world cultures, globalism is discovering itself in the assertion of rigid particularisms which show how little of social and cultural form is changed by the perceived logic of integration.

Mass forced migration is a child of the 'global era' and promises to be a prominent feature of world affairs. As Gill (1995: 94) observes, these 'unstoppable waves of migration' are unlikely to come to an end without positive intervention in the refugee-producing areas on a massive scale. 'The governments defending [sic] the regions of privilege', he comments, 'will be hard pressed to cope with or to contain such pressures' (Gill 1995: 95). Mass migration will continue to leave its mark on the most vulnerable regions, upon those coerced to leave them and upon societies that are desired places of asylum. As it does so a series of contradictions becomes more and apparent, for the same forces viewed as productive of world inclusiveness are making for a multilayered exclusion. In this respect, much of globetalk should be treated with great suspicion.

Notes

1. Some of the issues covered in this chapter were first raised at a seminar series organised by the Centre for New Ethnicities Research (CNER) at the University of East London and subsequently in the CNER's *Working Papers*. My thanks to Alice Bloch, Phil Cohen and Ray Kiely for their comments and criticisms at this stage.
2. There is a widespread view that population movements as a whole have taken on a global dimension. Among recent analyses, Collinson (1993: 4) writes of 'global causes and impacts of migration'. For Castles (1993: 17), 'The world is entering a new phase of population movements in which migration . . . can be fully understood only in a global context.' Overbeek (1993: 16) sees the increased pace of

forced migration as consistent with such changes, asserting that 'The rapid rise in the numbers of refugees [is] truly a global process.'

3. Such approaches have mesmerised the Western media. One review of global politics declared: 'The power of global capital in the form of transnational organisations and institutions has made a nonsense of national economic decision-making . . . Today this transformation is common wisdom' (*New Internationalist* 277: 9).
4. Harris's work is an interesting exception to the rule, being developed precisely in the context of the West's encounter with the Third World. This makes the conclusion of his study of migrant workers, *The New Untouchables*, more significant. Here, Harris invokes Hegel's Spirit of Reason as a synonym for the unfolding logic of the new world economic order (Harris 1995: 228). He draws his conclusion with full awareness of the implications; most other theorists operate unselfconsciously within fundamentally Eurocentric perspectives.
5. See Ruigrok and Van Tulder (1995), Hirst and Thompson (1996), Harman (1996) and Hoogvelt (1997). For an earlier, pioneering critique, see Gordon (1988).
6. See especially Hoogvelt (1997; ch. 6).
7. The phrasing is unfortunate: refugees are invariably described as moving in 'waves' or 'tides' – as dehumanised masses – and with an implicit threat that others are to be submerged by their presence.
8. There are signs that the relative tolerance shown in Third World host countries is beginning to break down as some refugee crises emerge on a massive scale. In February 1996 the Tanzanian government, for example, closed its borders to new migrations of Rwandans. Similarly, Liberian refugees were denied asylum in a series of West African states.
9. Interviews with Ethiopian refugees in Cairo, May 1995, and in London, April 1996.
10. This picture is confirmed by the writer's own observations of Sudanese refugees in various areas of the Middle East, for whom new means of communication have greatly heightened their focus on Return. As one Sudanese activist in Cairo commented: 'Now I am closer to my comrades in Khartoum than I was to my family in Darfur [in the west of Sudan] when I lived in Khartoum.' Interviews with Sudanese refugees in Cairo, December 1992.
11. Santel quotes Martin on refugees' perception of the successful migrations of Vietnamese and Chileans: 'if it worked for them why shouldn't it work for others?' (1993: 81).
12. In 1989, for example, asylum-seekers who had travelled for days from the conflict zones of Kurdistan to provincial airports in Turkey found themselves in London after a journey of less than four hours. By this means several thousand Kurds escaped the attentions of the Turkish state literally overnight.
13. Egyptian Islamist leader 'Omar Abd al-Rahman, for example, exiled (now imprisoned) in the United States, for several years used electronic means to lead a network of activists spread across Central Asia, the Middle East, Europe and North America. Saudi dissident Muhammed al-Masa'ari has similarly used Britain as a base from which to organise a network of supporters across the Gulf region. Masa'ari's Committee for the Defence of Legitimate Rights (CDLR) has concentrated its energies on faxing a weekly newsletter to Saudi Arabia, claiming a

readership of 300,000. CDLR activities have been described as 'unstoppable, in the communications age' (*Independent* 5 January 1996).

14. One striking example is recent access to satellite broadcasting from Turkey, which has given hundreds of thousands of Turkish and Kurdish refugees in Western Europe a live daily link with their regions of origin. No Turkish/ Kurdish refugee community centre in London, Stockholm, Hamburg or Paris is complete without its wide TV screen and Insat dish. Such developments can have important implications for the refugee relationship with 'host' societies – confirming the sense of distance from an unfamiliar culture – and for the visions of 'home', viewed daily as official news, 'light entertainment' and sport.
15. Interviews with refugees from the Horn of Africa, London, March and April 1996.
16. See Milward (1992) for an account of the dynamics of European integration and of regionalism which has implications for the idea of a globalising world economy.
17. Huntington's article, in *Foreign Affairs* (Summer 1993), has acted as the focus for a new Orientalism, depicting a coming global conflict between Western culture and various imagined Third World blocs, including 'Confucianism' and 'Islam'.
18. For an account of the death by drowning of an estimated 300 clandestine immigrants, see the *Observer* 2 March 1997.
19. In 1995, the European Council on Refugees and Exiles (Ecre) cited the case of a Somali family of six which was 'chain-deported' from Belgium to the Czech Republic, to Slovakia and eventually to Ukraine, where it could no longer be traced by UN support agencies (*Guardian* 23 February 1995).
20. Among the victims of this policy in 1995 were over 100 refugees from various Middle Eastern and Central Asian conflict zones who in March and April 1995 traversed Russia, Belarus and the Baltic states by train for several weeks. Of the ninety Iraqis, fourteen Afghans, four Palestinians and two Iranians more than half were children, a quarter of whom were under ten (*Observer* 9 April 1995). Victims of racketeers who had promised them entry to Sweden, the refugees had already travelled thousands of miles before being shunted back and forth across border crossings in an insane ritual: during one five-day period they were moved between Russia and Latvia twelve times while the authorities bickered over who should take responsibility.
21. Interviews with Ethiopian, Eritrean and Sudanese refugees in London, 1996.
22. According to one report, 'Head teachers, hospital administrators, housing and social security officials are to be trained and encouraged to identify suspected illegal immigrants and report them to the Home Office' (*Guardian* 19 July 1995). Even before these measures had become law, some employers had threatened their staff with 'spot checks' on their status and it seemed likely that up to 2 million people a year might be forced by employers to prove their identities with a passport or birth certificate (*Independent* 21 November 1995).

References

Arrighi, G. (1996) 'Whither the nation state?', *New Times* 20 January.

Axford, B. (1995) *The Global System: Economics, Politics and Culture*, Cambridge: Polity.

Balibar, E. (1991) 'Es Gibt keinen Staat in Europa: racism and politics in Europe today', *New Left Review* 186.

Castles, S. (1993) 'Migrations and minorities in Europe. Perspectives for the 1990s: eleven hypotheses', in J. Wrench and J. Solomos (eds) *Racism and Migration in Western Europe*, Oxford: Berg.

Cohen, R. and Joly, D. (1989) 'The "new" refugees of Europe', in R. Cohen and D. Joly (eds) *Reluctant Hosts: Europe and its Refugees*, Aldershot: Avebury.

Collinson, S. (1994) *Europe and International Migration*, London: Pinter.

—— (1996) *Shore to Shore: the Politics of Migration in Euro-Maghreb Relations*, London: RIIA.

Cooper, D. (1992) *Urban Refugees: Ethiopians and Eritreans in Cairo, Cairo Papers in Social Science*, vol. XIV, Monograph 2, Cairo: The American University in Cairo.

Cox, R. W. (1995) 'Critical political economy', in B. Hettne (ed.) *International Political Economy: Understanding Global Disorder*, London: Zed.

Delanty, G. (1995) *Inventing Europe: Idea, Identity, Reality*, Basingstoke: Macmillan.

Featherstone, M. (1992) 'Global culture: an introduction', in M. Featherstone (ed.) *Global Culture*, London: Sage.

Giddens, A. (1991) *Modernity and Self-Identity*, Cambridge: Polity.

Gill, S. (1995) 'Theorizing the interregnum: the double movement and global politics in the 1990s', in B. Hettne (ed.) *International Political Economy: Understanding Global Disorder*, London: Zed.

Gordon, D. (1998) 'The global economy', *New Left Review* 168.

Harman, C. (1996) 'Globalisation: a critique of the new orthodoxy', *International Socialism* 73.

Harris, N. (1995) *The New Untouchables: Immigration and the New World Order*, London: I. B. Tauris.

Harvey, D. (1989) *The Condition of Postmodernity*, Oxford: Blackwell.

Hirst, P. and Thompson, G. (1996) *Globalization in Question*, Cambridge: Polity.

Hoogvelt, A. (1997) *Globalisation and the Postcolonial World*, Basingstoke: Macmillan.

Huntington, S. (1993) 'The clash of civilizations?', *Foreign Affairs* 72: 3.

Jameson, F. (1991) *Postmodernism or the Cultural Logic of Late Capitalism*, London: Verso.

King, R. (1995) 'Migrations, globalization and place', in D. Massey and P. Jess (eds) *A Place in the World?* Oxford: Oxford University Press/Open University.

Mann, M. (1992) *The Sources of Political Power*, Cambridge: Polity.

Martiniello, M. (1994) 'Citizenship of the European Union: a critical view', in R. Baubock (ed.) *From Aliens to Citizens: Redefining the Status of Immigrants in Europe*, Aldershot: Avebury.

Milward, A. S. (1992) *The European Invention of the Nation State*, London: Routledge.

Ohmae, K. (1991) *The Borderless World*, London: Fontana.

—— (1995) *The End of the Nation-State: the Rise of Regional Economies*, New York: HarperCollins.

Overbeek, H. (1993) 'Towards a new international migration regime: globalisation, migration and the internationalisation of the state', in R. Miles and D. Thranhardt (eds) *Migration and European Integration*, London: Pinter.

Papademetriou, D. (1993) 'Confronting the challenge of transnational migration:

domestic and international responses', in OECD, *The Changing Course of International Migration*, Paris: OECD.

Robertson, R. (1992) *Globalization: Social Theory and Global Culture*, London: Routledge.

Ruigrok, W. and Van Tulder, L. (1995) *The Logic of International Restructuring*, London: Routledge.

Santel, B. (1993) 'Loss of control: the build-up of a European migration and asylum regime', in R. Miles and D. Thranhardt (eds) *Migration and European Integration*, London: Pinter.

Schwarz, B. (1996) 'The expansion and contraction of England', in B. Schwarz (ed.) *The Expansion of England*, London: Routledge.

Smith, A. (1992) 'National identity and the idea of European unity', *International Affairs* 68: 1.

Suhrke, A. (1997) 'Uncertain globalization: refugee movements in the second half of the twentieth century', in W. Gungwu (ed.) *Global History and Migrations*, Boulder, CO: Westview.

Tuitt, P. (1996) *False Images: the Law's Construction of the Refugee*, London: Pluto.

UNHCR (1995) *The State of the World's Refugees*, Oxford: Oxford University Press.

Weiner, M. (1997) 'The global migration crisis', in W. Gungwu (ed.) *Global History and Migrations*, Boulder, CO: Westview.

Widgren, J. (1993) 'Movements of refugees and asylum-seekers: recent trends in a comparative perspective', in OECD, *The Changing Course of International Migration*, Paris: OECD.

Zolberg, E. R., Suhrke, A. and Aguay, S. (1989) *Escape from Violence: the Refugee Crisis in the Developing World*, New York: Oxford University Press.

chapter 4

Global aspects of health and health policy in Third World countries

Maureen Larkin

chapter 4

Introduction

The health of populations is shaped by complex interactions between humans and their socio-economic, physical and cultural environments. The epidemiological profile (patterns of health and disease) of populations changes over time in intricate and uneven ways as the relationship to these environments change. This is exemplified in the major epidemiological transitions achieved in Western countries over the past 150 years which accompanied the socio-economic changes associated with the transition to an industrial capitalist society.

Over this period, major changes took place in patterns of health and disease and the life expectancy of populations. Significantly, major infectious diseases such as tuberculosis, and others such as the food- and water-borne diseases, which were the major causes of death in the nineteenth century, were wiped out, resulting in a substantial decline in mortality rates. As time went on, these diseases have come to be replaced by chronic conditions such as heart disease and cancers which are linked to changing lifestyles and increased longevity (McKeown 1976).

Whilst much debate surrounds the saliency of particular factors and processes involved in these transitions, research indicates that the key influences on health in this period were the preventive roles played by access to adequate nutrition, clean water and sanitation.

Medical science at this time lacked effective cures for the major causes of mortality. However, these changes cannot be assumed to have been a simple automatic outcome of industrial growth. Rather a variety of mediating processes of a political and ideological nature were at play to shape the developmental process and the accompanying health outcomes. Mobilising agents, ideologies and reform movements such as sanitary movements and trade unions and local government were an integral part of the change process exerting pressure on employers and the state for a more equitable distribution of economic growth (Szreter 1988). Moreover, improvements in health were not equally distributed throughout the population and differentials in health status between different socio-economic groups were marked and continue to be so up to the present day (*Black Report* 1980).

This brief synopsis suggests important similarities and contrasts which can be drawn between First and Third World countries in relation to health and development. In varying and uneven ways many of these countries are now undergoing forms of capitalist industrial development. Their health profiles are similarly distributed in uneven forms. In the poorer regions such as sub-Saharan Africa, parts of Latin America, Asia and the Middle East can be found health and disease profiles which are not dissimilar to those which prevailed in the developed world a century ago, with high mortality rates and low life expectancy (World Bank 1993). Infant and child mortality are particularly high and are linked to the complex interaction of infectious diseases with malnutrition and poverty.

Alongside this, in some countries such as the newly industrialising (NICs), and for the elites and middle classes throughout the Third World, are to be found patterns similar to those in the developed world today with increased life expectancy and a growing prevalence of chronic disease and disability (Phillips and Verhasselt 1994).

The picture is therefore a mixed one with old and new patterns of disease to be found telescoped together in the urban slums and rich suburbs of many Third World countries. It is estimated, for example, that infant mortality rates in slum areas can be three times as high as city averages in many cities (Basta in Phillips and Verhasselt 1994).

What this suggests is that the forms of development which are taking place, and how they are working themselves through, are highly complex and there are no easy parallels to be drawn between these and Western forms of development. In Britain capitalist industrial development combined with general social development to provide the basic prerequisites for health such as access to nutrition, clean water and sanitation facilities as well as reforms in the fields of housing, health and welfare services. But this was achieved over an extended period of time and under very different historic, economic and political conditions to those of the Third World today.

Capitalism developed gradually in Britain over several centuries, transforming it into an industrial society in the eighteenth and nineteenth centuries. Indeed it took over a hundred years to develop the necessary institutional reforms to provide for the health and welfare needs of the population. In contrast to this, the various forms of development taking place in the Third World today are considerably speeded up as modern technology and communications insert themselves into cultures and communities in complex and uneven ways. Over a matter of decades, what were formerly largely agricultural subsistence communities are rapidly being transformed by globalised forms of production and consumption practices (see Chapters 1, 2, 6 and 7).

This, combined with their very different historic backgrounds and the dominant position of Western countries in what are now global markets, means that the future direction and possibilities for the poorer countries is highly uncertain. As Chapter 1 makes clear, this uncertainty is intensified by the increasing tendency for dominant countries to use their market power and financial leverage to dictate the context and form of development which can take place.

The health problems and prospects within the poorer regions of the Third World are intricately caught up in these developmental processes. It is within and through the complex web of influences at play that health and health issues come to be defined, shaped and experienced. However, these processes can no longer be understood solely within the confines of particular countries or nation-states. As Western technologies, cultures and ideologies spread around the globe and insert themselves into the Third World, so increasingly development and health issues become caught up in and are shaped by these globalising influences. Linking development and health, this chapter will attempt to focus on the operation of some of these global processes and how they play themselves out with often negative or uncertain consequences for health.

The global context of health

Depending upon how the term global is used, there is a sense in which health in the Third World has long had a global dimension. There is an abundant literature which documents the health impacts of colonialism. These range from the direct effects of conquest and wars, the dissemination and spread of new diseases and epidemics, and the health impacts of the slave trade to the indirect effects of a colonial legacy in the form of a distorted distribution of resources, the dominance of Western medicine and the undermining of traditional communities and systems of healing (Doyal 1979).

Although the legacy of this period continues to cast shadows over the contemporary health context, today's global context presents a far more complicated scenario. Chapter 1 showed that the ending of colonialism and the attempts of newly independent states to develop along Western lines has worked through pressures and compromises over time to enmesh more and more populations within a modern global order which operates to shape and constrain options and choices for health as well as development. Global developmental processes today are working to create environments which for a majority of people present major challenges to health. The growth of poverty and the risks to health from industrial hazards and pesticide contamination are but one area of concern. Alongside this are the ideologies of consumerism and individualism which have increasingly become a way of life for growing numbers of people around the globe. The complex ways in which these are inserted into local cultures and their impact on social relationships and identities are not easy to grasp. As old and new cultures blend and mix, traditional forms of social relationships, social supports and patterns of sexual activity can be undermined or transformed and new lifestyles forged which can have uncertain implications for health. Feeding into this is the growing influence of Western medicine which, combining as it increasingly does with aspects of traditional therapeutic practices and the market power of global pharmaceutical TNCs, creates its own problems of inappropriate prescribing, drug misuse and abuse.

Policies and funding for health are also caught up in a variety of global processes and the activities and conflicting interests of a growing number of multilateral agencies such as the World Bank, the World Health Organisation (WHO), bilateral donor aid agencies and TNCs in the drug, food and alcohol industries. As well as these, there are the increasing numbers of international and national NGOs, consumer groups and social movements, all of which are active in attempting to shape the policy agenda (Walt 1994). In this sense health and health policies have become highly contested and these conflicts are increasingly played out at an international level through a variety of strategies and vocabularies which are not easy to penetrate.

Some of these selected areas and processes will now be examined, starting with the ways in which health and Western-driven development strategies have interacted over time but particularly from the 1980s and the advent of neoliberal, market-driven approaches to development.

Health and development policies

Throughout the 1950s and 1960s, policies for health as well as for development were largely based on the assumption that what had worked for the West was also best for the other two-thirds of the world population. Western-style economic development, supplemented by Western-style biomedically based health care services, were assumed eventually to bring about an amelioration of the health problems of all. This however has not been realised. While health gains in terms of reduced mortality rates and increased life expectancy were made, these gains were highly uneven with large disparities between countries and groups (Gray 1993). The failure of growth to trickle down, persistent poverty and growing inequalities in the distribution of resources forced a rethink both in the development field and in international health policy.

As indicated in Chapter 1, the development strategies of the global agencies such as the World Bank and the International Labour Organisation (ILO) changed in the direction of advocating policies for redistribution and meeting basic needs. Parallel to this came a reorientation of WHO policy away from the traditional high-tech, hospital-based approach which had characterised its policies for health in the developing world up to the 1970s. Catering as it largely did to the health needs of the urban-based middle classes, this approach was now considered inappropriate for meeting the health needs of the poorer populations. A more equitably based system of health care, which focused on meeting the basic health needs of populations in rural areas in particular, was advocated. This was known as the Primary Care Approach and sought to combine basic and accessible health services with the promotion of clean water supplies, sanitation and access to nutritional supplies (WHO 1978; and see pp. 105–6).

However, the development context within which Primary Care policies were to be implemented had by now moved on and development strategies had again changed. A combination of rising interest rates, falling commodity prices and rising indebtedness of Third World countries ushered in the World Bank/IMF's new neo-liberal development paradigm. As part of this strategy, structural adjustment policies (SAPs) were put in place in many countries which created major constraints on the implementation of WHO policies and exacerbated the already severe health problems which prevailed. As Chapter 1 showed, SAPs involved the removal of price controls, cuts in subsidies and rising food prices which worked to undermine food security and the nutritional status of large numbers of waged labourers in rural and especially urban areas. This, combined with labour market deregulation which worked to depress incomes and employment opportunities, meant that many households were unable to meet minimum food needs (Cornia *et al.* 1987). Cuts in state expenditure also put severe pressure on health budgets with resulting shortages of personnel, essential drugs and medical supplies, a decline in child immunisation programmes and the introduction of user charges (Asthana in Phillips and Verhasselt 1994).

In all, these policies have created serious adverse conditions for health. Whatever the long-term prospects for structurally adjusting countries, the short-term effects on the living standards of the poorer populations have resulted in a slowing down or worsening of health status in the regions most severely hit such as sub-Saharan Africa, parts of Latin America and the Middle East, where a deterioration in the nutritional status of children, increased incidence of infectious disease and raised infant and maternal mortality rates are reported (UNICEF 1995; Cornia *et al.* 1987).

Running alongside and compounding these problems has been the displacement of populations, violence and civil unrest which these dislocations have contributed to and which create their own physical and psychological threats to health. It is clear that prospects for health are now more than ever before complicated and constrained by the operation of development processes which to a large extent fall outside the control of local states and communities. Policies and mobilisations for health must therefore increasingly engage with multilateral development agencies such as the World Bank/IMF and their major funders (discussed further on pp. 104–5).

Expanding industrial activity

At the same time we can see how industrial activity is speeded up by intense competition within the global market economy and the activities of TNCs create altered patterns of work and new hazards for health. The weak position of many countries within the global economic order puts pressure on governments to create favourable environments for TNCs, and to an increasing extent for local investors. This, combined with developing countries' lack of experience and resources for the management and regulation of heavy industry, often results in a relaxation or absence of environmental controls, presenting serious health risks for workers and their families. These risks are found in unsafe plant design and production processes, lax safety standards and exposure of workers to hazardous materials and chemicals which are often banned or restricted in the developed countries (Cooper Weil *et al.* 1990). It is estimated, for example, that a manufacturing worker in Pakistan is approximately eight times more likely to die in an accident on the job than a worker in France (World Bank 1995).

In many countries, including those recently caught up in the rapid scramble for development such as the Philippines, Thailand, Indonesia and China, workers (especially women) work long hours for low levels of pay, often in cramped and poorly ventilated conditions. In Jakarta, for example, where many of the global names in sports shoe manufacture such as Reebok and NIKE are to be found, we read of young women earning 16p an hour and working in 90 degree heat. Work is contracted out by the TNCs who pass responsibility for terms and conditions to national governments. A work day can stretch to 18 hours and free trade unions are illegal (*Guardian* 3 December 1996).

Mechanisation and chemicalisation of agriculture

Other hazards are presented by the mechanisation and chemicalisation of agriculture as Western technologies and techniques are transported around the world through the activities of giant agribusinesses. There are a variety of health problems linked to irrigation schemes and resettlement of populations which are beyond the scope of this chapter (see Cooper Weil *et al.* 1990). Pesticides are, however, one important area where there is a growing concern about their misuse and abuse. The pesticide market is truly global and dominated by a small number of transnational corporations and the top ten companies control 73 per cent of the world market (Dinham 1993). Imports of pesticides to developing countries have shown a continuous increase over the past decade and a half. This is encouraged by pressure from the IMF to increase cash crops for export in order to service their debts.

Widespread use and abuse (though not adequately monitored) pose increasing problems of pesticide poisoning, pest resistance, environmental contamination and pesticide residues in food (Cooper Weil *et al.* 1990). The World Health Organisation estimates that more than 1 million workers across Latin America alone are poisoned every year and 10,000 die from exposure to chemicals (World Bank 1995). Aerial spraying carries particular hazards not only for operators but also for people and crops in non-target areas. For example, local practices of drying food stuffs such as couscous, spices, meat and fish on the roofs expose food to aerial spray drift. Health and safety issues are accentuated in these countries by a lack of hazard awareness both on the part of health professionals and of users, a lack of regulatory authorities, and problems surrounding labelling, language and high levels of illiteracy amongst the local population. Adequate washing facilities and protective clothing are unavailable and problems of disposal are exacerbated where contaminated drums are frequently reused for storing water and other food stuffs (Dinham 1993).

Health and the urban context

Other problems for health are thrown up by the rapid rates of urbanisation now being witnessed across the Third World. These are creating new and uncertain conditions which can have many health implications. Within the new urban environments, cultures blend and mix and social relationships are made and remade within the global context of modern cities. Individualistic values come to challenge, modify or replace the more traditional normative controls on identities and behaviours as migrants move from rural to urban areas and adapt to urban ways of life. International tourism and global media networks feed into this process, mediating and constructing new identities and realities.

This new social environment is conducive to altered patterns of sexual interaction which have particular significance for the development and spread of HIV/AIDS and other sexually transmitted diseases which present major challenges to health in the modern Third World. In some cities in sub-Saharan

Africa, it is estimated that as many as one in three adults are estimated to be infected with HIV (UNICEF 1994). As the traditional parameters governing sexual relationships are dismantled or weakened, more liberalised sexual behaviours can mean greater numbers and variety of casual sexual contacts and hence greater vulnerability to infection. A number of factors feed into this such as the growth of commercial sex and patterns of circular migration as separated partners move between rural and urban areas, thereby increasing the risks of infection transmission.

Underpinning and greatly aggravating these problems is the growth of poverty in urban areas which contributes to weakened resistance and greater vulnerability to infection. In some cities, it is estimated the poor can constitute up to 60 per cent of the population and are typically housed in urban slums and squatter settlements (Harpham in Phillips and Verhasselt 1994). This presents major threats to health in the form of overcrowding, contaminated water supplies and inadequate sanitation. These factors, combined with high levels of malnutrition, create major threats to health and life with high infant mortality rates from infectious diseases, particularly diarrhoea and respiratory infections (Harpham 1995).

Less researched but recognised as a growing health issue is mental health (World Bank 1993). Although major problems are implicated in how mental health problems are conceptualised in different cultures and reliable data is patchy, conditions known in the developed world to be associated with poor mental health status are abundantly prevalent in the urban environments of many countries. Amongst these are the social dislocation of migration, displacement from the labour force, erosion of the social supports of the rural family and community, poverty, stress and high levels of violence. What research exists points to mixed anxiety/depression disorders as the most common presenting mental health problems, with the highest prevalence amongst women (Harpham 1995).

However, no easy generalisations based on Western models of mental health or illness can be made without much greater understanding of the local historic and symbolic orders within which mental health concepts and problems are constructed and articulated. Local cultures combine with local circumstances to shape perceptions and the ability of people to cope. This complexity is well illustrated in the work of Sheper-Hughes on slum dwellers in north-east Brazil. Here we get a sense of what it means to live on a daily basis with chronic hunger, illness and death and the capacity of people through ritual and the pragmatics of everyday life to cope and survive in extremely adverse circumstances. The apparent indifference and stoicism towards infant death, displayed by the mothers in her study, is located by Sheper-Hughes within an ethnopsychology of emotion and the culturally specific constructions of self, personhood and society within an environment hostile to child survival and motherhood (Sheper-Hughes 1992).

Concepts of health and prescriptions for health are therefore not ideas which can uncritically be transported across cultural boundaries. As Western ideas and technologies of health are increasingly available around the globe, there is a need for greater understanding of how these are mediated by local socio-cultural circumstances. The globalised nature of Western medicine and how it is appropriated at a local level provides an interesting example in this respect.

The global context of pharmaceutical supplies

There is now a considerable body of research on the role of pharmaceutical TNCs in the production and promotion of Western medicine in Third World countries, involving problems of inflated prices, unethical marketing, drug dumping, the overuse and misuse of drugs and the growing problem of drug-resistant bacterial strains (Chetley 1986; Kanji and Hardon 1992; Melrose 1982). These problems are further confounded by the complex ways in which the operations of TNCs and Western medicine come to insert themselves into the institutional and cultural context of Third World countries creating new and often contradictory patterns of drug use. This expanding market in drugs is also working, as we shall see, to transform many traditional medicines into globalised commodity forms which, stripped of their therapeutic context, and packaged in standardised forms, have become available around the globe.

The market for Western drugs

Western medicine, first introduced into the Third World during the colonial era, was given official sanction by the newly independent states, most of whom modelled their health care systems on the West and thus provided an important market for transnational pharmaceutical companies (TNPCs). Associated with the former coloniser and with local elites who patronised it, Western medicine came to achieve a high status throughout these countries. The cost of drugs, however, placed a severe burden on the health budgets of the poorer countries (and continues to do so) and few outside the elites and the middle classes in the urban areas have easy access to modern pharmaceuticals.

With some exceptions, traditional medicines and systems of healing were, and continue to be, the main source of care for the mass of the people. This, however, did not stop a rapid expansion of a wide variety of drugs flooding the private marketplace where a lack of appropriate infrastructure by way of regulation and trained personnel created a chaotic system of drug distribution and use (Kanji and Hardon 1992). Useless and dangerous drugs were combined and dispensed often without prescription or information (except that supplied by the drug company) by a variety of untrained personnel from pharmacists to the local grocer or market vendor.

Although a variety of campaigns and oppositions from the 1970s has resulted in some curtailment of the activities of TNPCs, formidable obstacles remain outside of the public sector in securing a rationalisation of drug supplies in most countries (Chetley 1990). Large quantities of potentially hazardous drugs continue to circulate outside the control of health authorities and the private and informal sectors continue to be the main sources of drugs (Kanji and Hardon 1992).

Research around these problems has largely concerned itself with the conflicts and problematics surrounding the formulation and implementation of frameworks and initiatives for securing appropriate drug supply and use. Less attention has been given to the socio-cultural context in which drug supplies are appropriated at a local level. Without an appreciation of this context, it is difficult to see how effective regulation can be achieved.

Medicines and culture

Western medicines insert themselves into and are appropriated within complex cultural and healing systems at a local level. People buy, sell and consume medicines through culturally prescribed understandings of health and disease, concepts of causation and therapeutic relationships and regimes which endow them with meaning. A wide variety of traditional systems of healing and medicines are to be found throughout the Third World where traditional healers provide for the bulk of the health care needs of rural populations, especially in Asia and Africa (Phillips 1990). These practices run alongside and interact in complex ways with formal Western-style biomedical facilities and collaboration between the two is now actively encouraged by such bodies as the WHO.

Traditional systems range from the major classical traditions of China and India to a variety of religious or ethnically based systems to be found in many countries of the Third World (Phillips 1990). Such systems are based upon differing assumptions about cause and effect in health and illness as in the case of the classic humoral or equilibrium models, where a person's health is understood as a complex function of the balance between the individual and a variety of environmental forces and agents.

This can have implications for drug prescribing and drug use within the global context of the Third World where the mixing and overlap of traditional and modern forms of medicine and therapeutic practices are common. Bledsoe, for example, shows how traditional cultural assumptions about disease causation are ascribed to modern pharmaceuticals by the Mende people of Sierra Leone where red-coated tablets were sought out to replace blood loss regardless of their chemical properties. A preference for yellow tablets was noted in the case of the yellow urine associated with malaria. This relates to fighting fire with fire. Hot and cold are also further parameters within which the appropriateness of particular pills is judged. Injections, for example, are seen as cold and Bledsoe reports that women often applied hot compresses after such treatment, which had the effect of counteracting its therapeutic effect. Itinerant drug salesmen also prescribed Western drugs in accordance with traditional beliefs, quantity and frequency being related to the degree of perceived gravity of the disease or the strength of the person or even the season (Bledsoe in Geest 1988).

Western drugs are popular and are sought out and used alongside traditional therapies, sometimes in preference to them. Apart from their perceived therapeutic effectiveness, there are a multitude of reasons which account for this, not least of which is the ideology of West is best and the promise inherent in the promotion of Western pharmaceuticals of a pill for every ill. The quick fix offered by Western medicine may often be preferable to the complicated and lengthy practices surrounding the administration and use of many traditional therapies. Also the general trend towards individualism, and the weakening of kin and community relations as sources of identity and social support, create a context in which Western medicine may be more compatible with the new globalised lifestyles. As Western medicine treats only individual symptoms, and thereby discounts the more holistic and complicated involvement of kin and ritual, it may be perceived as more convenient and desirable within this context (Whyte in Geest 1988).

What is also evident is the way in which modern and traditional medicines are increasingly combined in the treatment and dispensing practices of a variety of indigenous practitioners whose survival has come to depend upon adaptation to the modernisation process and the increasing commercialisation of all forms of medicine and health care. In a variety of often inappropriate ways, traditional and modern medicines are combined, thereby offering both the reassurance and the familiarity of local traditional practice and the 'magic bullet' effect of the modern. Relying on shared cultural understanding and information gleaned from drug representatives and other sources, such traditional practitioners serve as important mediators between traditional and modern medicines.

Old and new practices are combined, for example, in the case of a cold, where advice to avoid cold foods and drinks and not to bathe run alongside a prescription for tetracycline or other broad spectrum antibiotic (Ferguson in Geest 1988). Other antibiotics used to treat stomach and diarrhoeal conditions were combined with advice regarding what foods and behaviours to avoid. Pills and tablets sold individually from bottles (partly for financial reasons) were noted to be dispensed in numbers which had significance within the local therapeutic cultures. It should be noted that such pharmacological syncretism both feeds into and is powerfully shaped by the promotion and marketing activities of the pharmaceutical TNCs whose representatives provide the main source of drug information and supplies for indigenous practitioners, who in turn have come to constitute an increasing clientele for them (Wolffers in Geest 1988).

At the same time, and at another level, the traditional and the modern are combined in ways that both compete with and reflect Western scientific traditions. Whyte, for example, argues that forms of traditional African medicines are increasingly being repackaged and re-presented by African governments and international agencies along Western lines (in Geest 1988). Denuded of their cultural symbols and the therapeutic relationships which underpinned their use, the emphasis is upon pharmacological properties and supporting scientific research.

These preparations are manufactured and packaged in standardised western forms by large and small firms and sold into global markets in the form of pills, cosmetics, powders and creams. The market success of these products could be argued to lie in their combined appeal to tradition and nature and their successful presentation within global contexts as 'modern' pharmaceutical products.

Such hybridisation of cultural forms and commercialisation makes orthodox notions of drug regulation and control highly problematic. Nevertheless, drugs and their regulation are an important issue at both international and national levels of health policy and constitute a major area of contention between a variety of conflicting interests.

The global policy context: oppositions and resistance

It is at the level of international policy-making, in particular, that we can glimpse the complex play of global processes and the many agencies and interests which are active in shaping health policy. This is particularly evident from the 1970s on as health policy became increasingly caught up in development policies and strategies, creating numerous conflicts and oppositions amongst the variety of agencies involved. Some strands of this process and the agencies involved will now be explored in relation to policies for drugs, primary care and some of the concerns surrounding aspects of women's health.

The World Health Organisation and policies for essential drugs

As a major multilateral organisation, the WHO is a major player in the global health policy arena. Through the World Health Assembly, which represents member states, the WHO sets the agenda for health formally in terms of standards and goals worldwide. And although its mandate does not stretch to implementation, which is a matter for local states, it does provide technical assistance and funding for health. What has become evident from the 1970s on is the way in which this organisation (like its many sister organisations in the United Nations) has been shaken up by tremors emanating from the failure of development to alleviate poverty in major regions of the Third World and the serious obstacles which this posed for health. A variety of agencies and challenges came into play in this period, from Third World countries to international consumer groups, women's groups and the growing strength of NGOs who worked to put pressure on the WHO for a more radical agenda for health.

Traditionally the WHO had been a rather conservative and sedate forum for research and information exchange rooted predominantly in a biomedical orientation. Resistances and challenges in this period which focused around the marketing practices and promotion of the major food and drug and tobacco

TNCs in the Third World worked to push the WHO towards a more radical agenda for health and the adoption of policies which brought them into direct conflict with the major TNCs and their governments (the best documented of which are the conflicts relating to the adoption of the International Code of Marketing of Breast Milk Substitutes, see Chetley 1986).

Intense confrontations have also surrounded the WHO's programme on essential drugs and attempts to control the marketing practices of pharmaceutical TNCs. Here we can see the major influences wielded by the pharmaceutical TNCs but also the major resistances which came into play around this issue.

An essential drugs policy was approved by the World Health Assembly in 1978 in response to pressures from a variety of groups to do something about the high drugs expenditure (in both developed and developing countries) and the proliferation of inappropriate, useless and dangerous drugs in the Third World. This list comprised 200-plus preparations which were deemed adequate to meet the pharmaceutical health needs of developing countries (Chetley 1990). The number of branded drugs available in some countries can run into several thousand.

This policy was fiercely opposed by the giant pharmaceutical TNCs supported by their national governments who saw it as a threat to their profitability. Major propaganda campaigns were mounted by the International Pharmaceutical Association which sought to discredit both the essential drugs programme and its supporters. These ranged from claims that it would discourage investment in research, undermine health care and promote third-rate drugs to charges that it would interfere with doctors' freedom to prescribe (Chowdhury 1995).

The motives and actions of organisations such as Health Action International (HAI) (an international coalition of NGOs and activists), who were major advocates of the policy and indeed wanted to see tougher regulation of the industry, were seen as a major threat and were presented by the industry as an international conspiracy with more than health on their minds. In the words of a Hoechst representative, 'the activists of HAI should be asked whether their true intention is not the overthrow of existing social and economic systems in favour of authoritarian regimes' (Chetley 1990: 72). And a Searle representative noted 'a world wide resurgence of consumer organization activity focused on the pharmaceutical industry and particularly multinationals' (*ibid.*: 72).

Obstruction and discrediting tactics were also employed at country level where countries attempted to develop rational drug policies. The better documented cases are Bangladesh and the Philippines. In 1982, the Bangladesh government legislated to regulate the supply of drugs across the private and public sector and to encourage its own domestic pharmaceutical industry. Opposition was marked from a combination of TNCs, their US and European home countries and the Bangladesh Medical Association in collaboration with some of the national press. Accusations of trade restrictions and reminders of the country's reliance on the West for investment and aid were used to exert pressure and secure the interests of the drug TNCs (Chowdhury 1995). The outcome of this experiment is still uncertain. After some considerable success, new political developments in the 1990s may see a modification or reversal of its fortunes (Reich 1994).

Drug TNCs, though challenged by these policies, have shown immense resilience and subtlety in protecting their interests, and deploying a variety of strategies to secure their dominance. These range from offers of cooperation in, for example, providing essential drugs at low cost to the development of their own voluntary code of marketing to regulate the industry (Chetley 1990). Relative to pressure groups such as HAI, the industry was also favourably placed within the policy arena to exert influence. Their representatives were and are regularly employed as advisers to the WHO and their international federation (IFPMA) has official NGO status within the organisation giving them ready access to expert committees and consultative meetings (Kanji and Hardon 1992). The industry has now largely come to accept the existence of essential drugs lists though within limitations, in that, even though a growing number of countries now have essential drug lists, these lists only apply to the public sector in the majority of countries and account for only a small percentage of total drug expenditures (*ibid.* 1992). In some countries, too, the list may not be mandatory and may not apply to the hospital sector which consumes most of the drugs budget.

Though these interests and oppositions continue to play themselves out, the international policy context is becoming further complicated by shifting funding arrangements and the growing influence of agencies other than the WHO in the health policy arena. The WHO's role and effectiveness in meeting the health needs of the poorer countries came to be questioned in the 1980s by donors and other UN bodies such as the World Bank, UNICEF and UNDP. With growing pressure on resources, donors started making greater demands for cost effectiveness in the use of funds and the WHO was subjected to criticism for its wastefulness, its bureaucracy and its inability to deliver the goods (Walt 1993). There was widespread criticism in the press for its lack of accountability, its remoteness and red tape (Godlee 1994; *Guardian* 19 May 1995; Walt 1993).

The World Bank

What we are now seeing is the growing influence of the World Bank as a major funder for health in the poorer countries and one which already eclipses that of the WHO and UNICEF. Its influence has grown steadily throughout the 1980s and 1990s as bank loans for health quadrupled (World Bank 1993). Its activities now reach into health service financing and delivery and the prioritisation of programmes along cost effective lines (World Bank 1993).

What the implications of this might be for the long-term health interests of these countries can only be speculative. Given its policies of a diminished role for the state and the privatisation and marketisation of health services as well as the leverage it can exert through its support for structural adjustment, the possibilities for equity-based health services are certainly in doubt. A major cause for concern must be its lack of accountability and the disproportionate weight accorded to the richer countries in the exercise of voting rights, as well as its current commitment to neo-liberal-type policies. Such constraints operate to render the influence on policy of popular interests and that of local states a great deal more problematic than it was a decade or so ago.

Loans and grants from the international community, where appropriately allocated, can play an important part in enabling poorer countries to provide for the health needs of its populations. Whilst it is arguable how far dedicated health funds as opposed to other broader development funds can adequately meet the pressing health problems rooted in poverty, there is certainly an argument to be made that some forms of health funding can be more appropriate than others.

With health budgets in low-income countries, now increasingly dependent upon donor aid, which for some countries can constitute more than 50 per cent of all health expenditures (World Bank 1993), there are concerns that donor aid in the area of health may be more reflective of the interests of the donors rather than the long-term health needs of recipient populations (Walt 1994). Desperate as many of the poorer countries are for funds, they are not favourably placed to dictate what or which area of health ought to get funding.

Primary care

We can see how these constraints have worked to shape policy in the field of primary care. As outlined above, comprehensive primary care became the WHO's major policy initiative in the 1970s. It was an attempt to go beyond the narrow medically oriented approach of past policies and to stimulate the development of a broad-based strategy for health which combined basic health needs (particularly for rural communities) with community development and intersectoral collaboration in the fields of education, food, water and sanitation provision (WHO 1978).

However, the deterioration of the economic and social conditions of the 1980s saw many twists and turns in the funding and implementation of these policies, with the result that primary care has now come to mean a variety of different initiatives or none at all in many of the poorer countries. The interests of donors have played no small part in the fortunes of this policy.

In the context of the 1980s, comprehensive forms of primary care came to be seen by many donors as too ambitious, expensive and uncertain in outcome. Instead they came to favour more cost-effective, individual and largely biomedical technological initiatives, which could demonstrate quantifiable and measurable outcomes. The better known of these are growth monitoring, immunisation, the promotion of breast feeding, oral rehydration and family planning (Phillips 1990).

This reorientation of policy towards what has come to be called selective primary health care (SPHC) was reflected in debates between scholars about the effectiveness of the different strategies and the conflicts between the WHO and its sister organisation UNICEF, who favoured the selective approach. Critics of the selective approach argue that while such interventions can save lives in the short term, they do not address the underlying problems of poverty, malnutrition and the absence of basic water and sanitation amenities. As well as a variety of other problems such as lack of continuity and integration of services, it is also accused of being professionally and medically dominated, thus continuing a pattern of dependency on western expertise (Gish 1979; Bannerji 1988).

Walt's research on the financial plight of the WHO highlights the way in which donors such as the World Bank and bilateral donor governments are gaining increasing leverage over the WHO's policy making. In bypassing the WHO's regular budget (frozen in real terms in the 1980s) and putting their money into extra-budgetary funds targeted towards specific programmes, they are in a much stronger position to shape policy in the direction of their own interests (Walt 1993). Extra-budgetary funds, which now constitute over half of the WHO's total income, are used for special projects such as the global AIDS programme or the immunisation campaign, amongst others. This means that the voices of the WHO's member countries from the developing world will carry less weight in deciding the direction of funding (Godlee 1995).

Bilateral agencies (country to country) are also increasingly channelling their funds through the World Bank, the IMF and the NGOs, all of which works to weaken the influence of local states in the policy-making arena (Walt 1993).

Resistance and compromise

What then are the prospects for those attempting to influence health policy in a more equitable and sustainable direction? As examples above indicate, there has been a major growth of opposition groups and movements since the 1970s which have presented challenges to the dominant policy players. The ability of these groups to sustain their campaigns is considerably strengthened by the growth and coordination of similar mobilisations around the world in related fields such as environmental, human rights and women's issues. However, it is extremely difficult to penetrate the international web of interests and alliances at play and how activities and issues play themselves out through what are complex forms of compromise, incorporation and marginalisation.

Women's health issues

Women's movements and women's health issues are a telling example of the problems and prospects which confront those seeking to influence policy in a more equitable direction. Women's status and women's health are now major issues on the international policy agendas of the major developmental and health agencies. Pressure from the international women's movement and grassroots campaigns, as well as the recognition by the development establishment of the potential contribution of women to the 'development' effort, has meant that no serious policy for health or development can any longer ignore a woman or gender dimension. Grassroots campaigns around development and women's health are now a major political reality especially in Third World countries and international collaboration has continued to grow since the 1975/85 UN Decade for the Advancement of Women. (For an overview of the important health issues related to women in the Third World, see Smyke 1991; WHO 1995.)

However, although all groups, including the major donors, claim to prioritise women's interests, there are many different perspectives and strategies deployed both nationally and internationally in addressing them. Cleeves Mosse identifies two broad tendencies. On the one hand are the more needs-based welfarist approaches most favoured by the major donors and national governments and which attempt to couple improving the status of women through education and income generation with provision for mother and child health and family planning.

On the other hand are the more radical and feminist-based approaches which locate women's health issues within the broader context of unequal gender relations and sees the necessity for a more bottom-up grassroots approach of empowering women. Such approaches are less likely to gain more than a rhetorical helping hand from governments or major donors (Cleeves Mosse 1993).

These positions and the conflicts which they engender are often glimpsed at international conferences where they become caught up in a variety of political and cultural conflicts which cut across gender divisions and work to reconstitute them in often contradictory and uncertain ways. This was seen in the UN Cairo Conference on Population 1994 and the Beijing World Conference on Women 1995 where a discourse of needs versus rights, especially around reproduction and sexuality, became entangled in the global ideologies of the Vatican and Islam (Petchesky 1995; *Independent* 5 September 1994).

There is, however, no doubt that women's movements have been and continue to be a major force in instilling a gender awareness into the policies and practices of agencies in the fields of development and health at both international and local levels and in the mobilisation of women around these issues. The major campaigns mounted around contraceptive issues and the challenges to the ideology of population control are notable examples of the critical role played by women as arbiters of women's health (Hartman 1987).

It is nevertheless unclear how far working through official bodies such as the UN agencies or local states can amount to fundamental challenges to entrenched gender interests and the many technological, political and religious forces which go to shape policy in this area. The issue of reproductive health, for example, highlights the many conflicts and compromises which gender issues throw up and the uneasy alliances which are often forged between a variety of conflicting interests.

Reproductive health issues

Women's reproductive health is a major health issue for the poorer countries, not least because of continuing high rates of maternal mortality, especially in

South Asia and Africa where women are at least a hundred times more likely to die in a single pregnancy than a woman from the developed world (Doyal 1995).

While it would appear that major progress has been made within the international policy community in transforming the discourses which frame the issue of reproductive health, outcomes in terms of policy implementation are a great deal more uncertain.

Over the last decade or so policies in this area have redefined and remade as pressure from women's groups, combined with policy failure in the area of maternal health, worked to push thinking beyond previously narrow concerns with family planning and mother and child health. These programmes, in which women were targeted largely in their capacities as bearers and rearers of children, have come to be modified or superseded by strategies which claim to locate maternal health within a broader woman-centred approach. Women's needs across the lifespan and in their productive as well as reproductive roles need to be addressed (Sweetman 1994).

Major discursive transformations have taken place particularly at the level of the international women's movement, where reproductive health has come to be defined in terms of reproductive rights, the empowerment of women and gender equality. Interestingly, too, population and family planning groups, who have come under increasing pressure from the Vatican and fundamentalist groups, also seem to have shifted their position, and in forging alliances with women's NGOs have come to embrace the new vocabulary of women's rights. How these latest conference proceedings work themselves through is another matter and certainly there are women's groups both North and South who are wary of such alliances and see the new discourse as nothing more than population control with a feminist face (Petchesky 1995).

It is notable that for some time the major development agencies such as the World Bank and population groups have deployed a status of women discourse in their analysis and programmes for development and population control. As research in the 1970s and 1980s came to link development, population and the status of women, these agencies came to recast their programmes in line with improving the status of women. Thus for the World Bank 'enhancing the status of women is of critical importance in strengthening the demand for smaller families' (World Bank 1986, quoted in Sweetman 1994). And, as Hartman demonstrates, anti-poverty programmes, education and income-generation projects have been promoted by population agencies, running alongside family planning programmes (Hartman 1987).

Richter's (1996) account of the recent and ongoing struggles surrounding immunological birth control methods (anti-fertility vaccines) demonstrates the immense versatility and subtlety of agencies such as The Population Council and others engaged in this research in their attempts to coopt or marginalise opposition from women's groups; from discrediting their objections as un-scientific and questioning their representativeness, to framing contraceptive research in terms of enhancing women's choice. There is also the fine line which women's movements must thread between the interests of the church and pro-life groups and the interests of women in many countries who lack access to any form of safe contraception (Richter 1996). All of this works to create potential cleavages and opportunities for the cooption of a woman-centred agenda.

Action at a local level

Changing vocabularies and policy formulation at the international levels notwithstanding, it is a great deal more difficult to assess their impact at a local level where a variety of officially sanctioned initiatives and informal women's projects are to be found. Thus, expanded women's health programmes emanating from the WHO and the World Bank, and which seek to integrate family planning with a broader range of services, have certainly been put in place in many areas, and there are a variety of more informal community-based women's health projects which are clearly based on feminist principles (Sweetman 1994; Doyal 1995).

However, many of the more formally based programmes have been criticised for being no more than add-on initiatives to existing programmes, such as screening and counselling, within the context of the same personnel and often the same clinics. Wider issues such as those identified above, as well as the material and cultural constraints which shape women's health chances, do not appear to be addressed in any significant sense.

There are also concerns about the high profile accorded to family planning within some of these programmes. Smyth, for example, argues that in the case of Indonesia's Safe Motherhood Campaign (a WHO/World Bank initiative) resources are disproportionately allocated in favour of family planning activities (Smyth in Sweetman 1994). There is clearly a need for a great deal more in-depth research in this area in order to clarify how these different interests work to shape policy at both the international and national levels. There can be no doubt but that the major players identified, such as population groups, donor aid and research agencies, form a powerful coalition of interests with both the financial and technical resources to potentially coopt their opposition. Women's groups have however shown a marked resilience in maintaining a critical scrutiny of policies and research in this area and in ensuring that women's interests are kept on the international policy agenda.

Conclusion

While clearly all the health problems of the poorer Third World cannot be explained away as a result of globalisation, it is nevertheless the case that health is intimately caught up in this process. Global processes which are shaping the development process are also both directly and indirectly shaping prospects for health. We have seen how current development strategies are working to create environments of poverty and health risks for large numbers of people. Means of livelihood and ways of life are being transformed as new production techniques, increasing urbanisation and cultures of modernity are inserted into traditional settings, creating new hazards and uncertainties for health.

At the same time, the ability to affect decision making in the procurement and allocation of resources for health appears to be increasingly displaced from local and national communities and has come to be concentrated in the hands of supranational bodies such as United Nations agencies and the variety of technical and commercial interests which patrol the international policy arena. The role of the World Bank, donor aid agencies and drug TNCs are a case in point. How far opposition groups active at international levels can be effective in challenging these dominant interests is highly problematic. There is no doubt that the globalisation process has created new sites of resistance and has brought together a variety of overlapping spheres of interest in the areas of health, development, environmental and women's groups. Local, regional and international networks have been created as communications technology and international travel have made contact and coordination a great deal more accessible. Global media networks have also worked to create a greater general awareness of and receptivity to Third World issues.

However, as demonstrated above, the dangers of dilution and cooption by dominant interests are an ever-present threat given the heterogenous nature of this opposition. This is compounded by the fact that global–local links, essential for effective mobilisation, are also often difficult to forge and get caught up in contradictions emanating from the particularistic interests of religion, ethnicity, class and gender. Without effective mobilisation of communities, radical international agendas are no more than fine rhetoric. Prospects for health therefore depend upon overcoming these obstacles and building alliances which can effectively challenge current policies in both development and health.

References

Bannerji, A. (1988) 'Hidden menace of the universal child immunization program', *International Journal of Health Services* 18(2): 293–9.

The Black Report and The Health Divide (1988), in P. Townsend and N. Davidson (eds) *Inequalities in Health*, London: Penguin.

Chetley, A. (1986) *The Politics of Baby Foods*, London: Frances Pinter.

—— (1990) *A Healthy Business: World Health and the Pharmaceutical Industry*, London: Zed Books.

Chowdhury, Z. (1995) *The Politics of Essential Drugs*, London: Zed.

Cleeves Mosse, J. (1993) *Half the World, Half a Chance*, Oxford: Oxfam.

Cooper Weil, D., Alicbusan, A., Wilson, J., Reich, M. and Bradley, D. (1990) *The Impact of Development Policies on Health*, Geneva: WHO.

Cornia, G., Jolly, R. and Stewart, F. (1987) *Adjustment with a Human Face*, Oxford: Oxford University Press.

Dinham, B. (1993) *The Pesticide Hazard*, London: Zed.

Doyal, L. (1979) *The Political Economy of Health*, London: Pluto.

—— (1995) *What Makes Women Sick*, London: Macmillan.

Geest, S. V. D. (1988) *The Context of Medicines in Developing Countries*, Amsterdam: Kluwer.

Gish, O. (1979) 'The political economy of primary care', *Social Science and Medicine* 13c: 203–11.

Godlee, F. (1994) 'WHO in crisis', *British Medical Journal* 309: 1424–8.

—— (1995) 'WHO's special programs', *British Medical Journal* 310: 178–82.

Gray, A. (1993) *World Health and Disease*, Milton Keynes: Open University Press.

Harpham, T. (ed.) (1995) *Urbanization and Mental Health in Developing Countries*, New York: Avebury.

Hartman, B. (1987) *Reproductive Rights and Wrongs: The Global Politics of Population Control and Contraceptive Choice*, New York: Harper Row.

Kanji, N. and Hardon, A. (1992) *Drugs Policy in Developing Countries*, London: Zed.

McKeown, T. (1976) *The Modern Rise of Populations*, London: Edward Arnold.

Melrose, D. (1982) *Bitter Pills: Medicines and the Third World Poor*, Oxford: Oxfam.

Petchesky, R. (1995) 'From population control to reproductive rights', *Reproductive Health Rights* 6, November.

Phillips, D. (1990) *Health and Health Care in the Third World*, London: Longman.

Phillips, D. and Verhasselt, R. (1994) *Health and Development*, London: Routledge.

Reich, M. (1994) 'Bangladesh pharmaceutical policies and politics', *Health Policy and Planning* 9(2): 130–43.

Richter, J. (1996) *Vaccination against Pregnancy: Miracle or Menace*, London: Zed.

Sanders, D. and Carver, R. (1985) *The Struggle for Health*, London: Macmillan.

Sheper-Hughes, N. (1992) *Death Without Weeping*, Los Angeles: University of California Press.

Smyke, P. (1991) *Women and Health*, London: Zed.

Sweetman, C. (ed.) (1994) *Population and Reproductive Rights*, Oxford: Oxfam.

Szreter, S. (1988) 'The importance of social intervention in Britain's mortality decline', in B. Davey, A. Gray and J. Seale (eds) (1995) *Health and Disease*, Milton Keynes: Open University Press.

Turner, B. (1996) *Medical Power and Social Knowledge*, London: Sage.

UNICEF (1994) *The State of the World's Children*, New York: UNICEF.

—— (1995) *The State of the World's Children*, New York: UNICEF.

Walt, J. (1993) 'WHO under stress: implications for health policy', *Health Policy* 24: 125–44.

—— (1994) *Health Policy*, London: Zed.

World Bank (1993) *World Development Report: Investing in Health*, Oxford: Oxford University Press.

—— (1995) *World Development Report: Workers in an Integrating World*, Oxford: Oxford University Press.

World Health Organisation (1978) 'Alma Ata 1978', in *Primary Health Care: Health for All*, Series No. 1, Geneva: WHO.

—— (1995) *Women and Health: United Nations 4th Conference on Women*, Special issue, Geneva: WHO.

chapter 5

The Biodiversity Convention and global sustainable development

Suzanne Biggs

This chapter aims to evaluate the impact on developing countries of current processes of globalisation, through an examination of contemporary approaches to the environment and development. These will be examined within the context of recent scientific and technological developments, with specific reference to biotechnology.

Introduction

As the twenty-first century approaches, concerned environmentalists are seeking to remodel macro development processes at the global level. In 1992, under the auspices of the United Nations (UN) at Rio, Brazil, an assembly of people, with diverse and often contradictory agendas, gathered from across the globe to discuss and ratify five key environmental framework conventions and statements intended to enable long-term sustainable development. Heads of state from the most powerful developed countries (G7) and developing countries (G77) attended. Representatives from the key transnational corporations (TNCs), the scientific community, environmental movements and NGOs were present.

Central to this event was the protection of the contemporary environment of 'planet earth' through the promotion at the global level of sustainable development. This was to be achieved through supranational action under the auspices of the United Nations. Global environmental politics had finally arrived when in 1992 this UN conference on Environment and Development (UNCED) proclaimed the environment (along with peacekeeping and economic development) as one of the three central pillars of world politics.

In 1983, responding to increasing concern about the environment, the UN authorised Gro Harlem Brundtland, then prime minister of Norway, to chair a commission charged with formulating 'a global agenda for change'. The commission identified an urgent need for decisive political action to put global development on a sustainable path for the twenty-first century and to begin to manage 'environmental resources to ensure both sustainable progress and human survival'. The World Commission on Environment and Development claimed that

> technology and social organisation can both be managed and improved to make way for a new era of economic growth . . . that humanity has the ability to make development sustainable . . . through a process of change in which the exploitation of resources, the direction of investments, the orientation of technological development, and institutional change are made consistent with future as well as present needs . . . which needs . . . include meeting the essential needs of the world's poor through sustained growth . . . and effective citizen participation.
>
> (World Commission on Environment and Development 1987: 8)

Rio initiated an ongoing process of international agreements concerning the environment and development brokered under the auspices of the United

Nations. The official agenda of Rio was the ratification of five agreements of which the Convention on Biological Diversity (CBD) to protect species and ecosystems and establish terms for accessing biological resources and relevant technologies was one.

The Convention on Biodiversity

Biodiversity (biological diversity) is the word currently used to describe all living organisms, their genetic makeup and the communities they form. Genetic diversity describes the variation of genes within a species; species diversity describes the variety of species within a region; and ecodiversity refers to the number and distribution of ecosystems. Global theory views planet earth as an integrated and interdependent ecosystem. Concern has been growing amongst scientists, policy makers and the public about the accelerating loss of biodiversity resulting from human impact. (Of the estimated 30 million species on earth, only 1.4 million have been described and United Nations Environmental Program (UNEP) (1992) estimates another quarter of these face extinction.) Loss of biological resources is said to threaten our physical existence, our food supplies, sources of medicines, energy and raw materials.

The Global Biodiversity Strategy (1992) identified fundamental causes of biodiversity loss. In 1995 a similar analysis was put forward when UNEP (1995: 3) cited the causes as being demographic, economic, institutional and technological. Loss, fragmentation, degradation of natural habitats were said to be the result of increasing demand for, and exploitation of, biological resources; the conversion of natural habitats to other uses and the growth in urbanisation and tourism; international trade. Inappropriate technologies were particularly heavily implicated in 'bio-loss' through food and livestock production and the failure of the market to recognise the true value of biological diversity, particularly at the local level. UNEP's understanding of the causes of the loss of biodiversity indicates critical concern amongst some environmental scientists about the direction of the current development process. The CBD was innovative in its attempt to give comprehensive and global expression to concerns about biodiversity and to further sustainable development by bringing together previously separate initiatives through three key objectives: the conservation of biological diversity, the sustainable use of this diversity and the fair and equitable sharing of the benefits arising out of the utilisation of genetic resources.

The global rhetoric of the 1992 UN conference and conventions suggests that environmental problems are global and amenable to global solutions through sustainable development. My overall contention is that the agenda agreed by participants to the CBD does not represent a suitable global framework for change which will effectively protect biodiversity at the global level, in accordance with the new paradigm of sustainable development. Moreover, in the context of recent developments in biotechnology, I shall seek to establish that this convention, far from being a global initiative capable of furthering sustainable development by accommodating the concerns and needs of the developing world as well as the developed world and operating equally in the interests of both, was conceived far too narrowly within the European tradition of conservation and emphasised environmental concerns at the expense of economic and development needs. Whatever its good intentions and possible benefits, the conservation framework of the CBD carries the burden of European colonial history and uneven development (discussed in Chapters 1 and 2).

How likely is it that an essentially environmental initiative and agreement will be able to create partnerships from within the uneven, disharmonious pattern of development which UNEP cites as a causal factor in environmental destruction? Are these likely to be partnerships which overcome poverty and stimulate more equitable growth at the same time as preserving the ecological base for sustainable development?

I shall argue that the cultural and institutional framework that the conservation movement has so far developed is too limited to be able to respond to the concerns raised about the contemporary economic development process in relation to biodiversity. At Rio it could not even ensure that genetic resources, held *ex situ* in the International Agricultural Research Centres, came under global governance. Discussions concerned with the transfer of technologies, particularly new biotechnologies, from developed to developing countries, were considered beyond the mandate of this convention. Sustainable development is unlikely to take off at the global level without the developing countries' full participation in the new biotechnologies. The latter part of my analysis therefore considers the corporate, institutional, economic and financial structures surrounding the development of new biotechnologies and considers whether these are likely to encourage the innovation of accessible and appropriate biotechnologies, particularly in the developing countries. Finally, I discuss how significant components of biodiversity could be privatised by TNCs if the World Trade Organisation (WTO) succumbs to pressure to extend and intensify patents and Trade Related Intellectual Property Rights (TRIPs). Issues related to economic development need to be on the agenda of a convention that intends to secure biodiversity and enable sustainable development at the global level, particularly as biotechnologies are likely to proliferate in the third millennium. The CBD is currently a very weak instrument in the face of other global actors and uneven economic development.

Northern interests in Southern locations

Both developed and developing countries were at the CBD negotiating table, the former to negotiate their interest in conservation, the latter to negotiate their interest in development. A central factor which precipitated some Northern actors, particularly the TNCs and their national governments from the developed world, to participate in the negotiations, is the location of *in situ* (in the wild or on farm) biodiversity and its accessibility. The industrialised North cannot compete with the less industrialised countries around the tropics in terms of assets in biodiversity. Many of the developed world's medicines and foods derive from and are dependent on continuous access to this biodiversity.

The location of biodiversity has recently become of increasing importance. Although the intensification of pharmaceutical and pesticide activities post Second World War in developed countries began to raise concern in some developing countries because of excessive profits and lack of technology transfer, research and development (R&D) in plant material for foods remained very much in the public laboratories and academia of the developed world and was to some extent accessible to developing countries. It is only with the recent development of the new biotechnologies and the increasing pressure from the TNCs for extended patents on their plant products that the developing countries have decided their interests might best be served by asserting sovereignty over the biodiversity within their borders. They hoped to use 'their' biodiversity to negotiate access to biotechnologies and recoup some of the profits generated from the food and medical technologies.

Biodiversity, drugs and the pharmaceutical interest

Access to biodiversity is critically important to the pharmaceutical industry, not only in the search for new products where plants form the basis of most of the medicines, but also to meet the increasing demand for natural medicines, raw and processed, in the developed world. UNEP estimated in its most recent biodiversity assessment that in 1993 about '80% of the 150 top prescription drugs used in the US were either synthetic compounds modelled on natural products, semi-synthetic compounds derived from natural products, or in a few cases natural products' (UNEP 1995: 14).

In 1990 the annual world market value of pharmaceuticals derived from medicinal plants and herbal drugs was estimated at US$43 billion. It is expected to reach US$47 billion by the year 2000 (*Multinational Monitor*, June 1994, cited in CIIR 1993: 8; Tolba *et al.* 1993; UNEP 1995).

Perhaps best known are the drugs used to treat childhood leukaemia which were developed through capital-intensive R&D by the American-based TNC Eli Lilley using the rare genetic trait of the rosy periwinkle found in Madagascar. Eli Lilley's sales of these drugs amount to approximately US$100 million per year worldwide but Madagascar has received nothing for the use of this resource (Miller 1995: 113). Many other traditional medicinal plants have similarly been taken and used to synthesise chemical substitutes without payment.

Plants can be screened for specific genes and genetic engineering can speed up the process of identifying new materials and developing new products. Tissue culture, recombinant DNA techniques and enzyme technologies are some of the new techniques being used to develop new crops with new characteristics, new vaccines and new sources of industrial components (Avramovic 1996: 9; CIIR 1993:18). Developing countries are now determined to utilise these revolutionary techniques themselves, using their local plant genetic resources. They have become unwilling to allow TNCs to corner the market in the developed world without adequate recompense. And they are extremely anxious to avoid TNCs cornering their national markets (Martin 1995: 6). In short, developing countries are no longer content to be left out of this process, and are intent on getting money for biodiversity and access to technology.

The contested agenda: global conservation or economic advantage

The negotiations for the CBD got under way in 1987 when UNEP called upon governments to examine the possibility of establishing an international legal instrument on the conservation and sustainable use of biodiversity. The Convention was drafted by an intergovernmental negotiating committee during the period 1988–92. This global vision of common interest in the conservation of the world's biodiversity was immediately challenged by long-standing interests. Two ongoing and highly significant conflicts between developed and developing countries occupied centre stage both during the negotiations and at the Convention, and they are still being fought over at post-Convention meetings – Conference of Parties meetings (COPs).

The first concerned ownership of and access to genetic resources and the second was over access to the new biotechnologies and extensive patents. Although many of the conveners of the Convention, particularly the Northern NGOs, had conservation as their focus, the industrialised interests of the North (TNCs backed by national governments) were at the negotiating table as much to try to access biodiversity on their own terms as to ensure conservation. Developing countries, on the other hand, used the negotiations to assert their intensifying grievances concerning the free use of their increasingly valuable genetic resources to develop pharmaceutical and food biotechnologies in the private sector, in companies predominantly located in the industrialised countries. Developing countries saw their interests best served through asserting sovereignty over their plant genetic resources.

Technology transfer: biotechnology as well as biodiversity

Early on in the negotiations developing countries threatened to withhold participation in the Convention unless measures were included to provide 'preferential and non-commercial' access to the biotechnologies that they needed to conserve and exploit their own resources. Developing countries insisted that the conference move beyond concern with the biological aspects of biodiversity and address socio-economic issues. Northern and Southern political and economic interests related to industrial development and technology were at stake.

The inclusion of biotechnology challenged the cultural frameworks of what were to many actors and the general public ostensibly conservation negotiations. Post-convention some NGOs accused the developing countries of 'perverting the cultural meaning of biodiversity from an aesthetic, moral understanding of nature linked to conservation, to a technological and economic concern' and of transforming 'the diversity of animal and plant life into a natural resource to be exploited and manipulated through modern science and its understanding of genetic codes' (Chatterjee and Finger 1994: 42). But while some of the conveners could only think of conservation as an aesthetic and ethical concern the environmental scientists working with UNEP appear to have accepted the economic agenda as the only possible way to enable the conservation of biodiversity. In order to understand the deep mistrust between developed and developing countries it is useful to explore two things: the conjuncture of biodiversity with biotechnology, and the history of conservation initiatives.

Biodiversity, biotechnology and discourses on the natural environment

The increasingly vociferous concern about the environment in the late twentieth century has been paralleled by new scientific knowledge and technological innovations. Just as new scientific insights can lead to the development of new technologies and industries, they also generate new understanding and knowledge of our world. Knowledge and language are intimately connected. The unravelling of the structure of DNA by Western scientists has changed our understanding of and relationship to natural life forms. Scientific knowledge is challenging discourses about nature, 'our world views' and cultural frameworks, as well as spatial and temporal relationships both within human society and between human society and the natural environment. The global spread of this latest universalising scientific knowledge has the potential for reducing all of nature to its component parts. Western science by unravelling the structure of DNA has moved deeper into the conquest and control of nature.

Contemporary Western science has through its increasing ability to manipulate DNA reduced nature to a series of genetic components which can be isolated and recombined. The articulation of this new language of biodiversity is concomitant with the new biotechnologies which can isolate genes from an organism, manipulate them in the laboratory and insert them stably in another organism. Nature is no longer a process embedded in space and time expressing itself in natural living species through the process of evolution taking place over time, within spatially delineated ecosystems. The component parts of nature can be disembedded and their relationship to space and time appear likely to be overcome.

This has brought the Western ontological understanding of nature into question. Profound insights and fearful responsibilities attend the increasing ability to manipulate DNA. Concerned people are struggling with both the global and local implications for humanity and nature. Scientific and popular terminology about nature has also changed. Biodiversity is an all-encompassing term which delineates a global, planetary phenomenon. It is an abstract and universalising terminology, encompassing not only species but also ecosystems and genes. 'The notion of biodiversity takes for granted the idea that all living beings are composed of related genes and that there can be a measure of genetic variability which is applicable across the globe' (Yearley in Anderson *et al.* 1995: 226). Biodiversity signifies a new understanding of all life forms 'represent(ing) the very foundation of human existence' (UNEP 1995: ii). Biodiversity is the essential ingredient of biotechnology which can recombine genetic traits in the DNA and use them to create genetically changed life forms. Biodiversity and biotechnology are intimately connected.

Technology is a defining entity and at different historical periods it enables different modes/ways for humanity of relating to the environment and to each other as individuals and as groups. It enables different social formations, different societies, different ways of being on the planet. However, technology itself emerges within and from different social structures. Industrial capitalism has been paramount in delivering new technologies (Rosenberg and Birdzell 1990). Just as our contemporary understandings of nature based in Western scientific knowledge have emerged in conjunction with new biotechnologies so these innovations are being developed within a capitalistic economic order. And as a result certain economic and political concerns have increasingly focused on biodiversity, investing it with economic value.

It is hardly surprising therefore that the language of capitalism has swiftly become an essential part of the attempts to describe the paramount importance of the natural world. Current and future generations' survival depends on the long-term sustainability of earth's 'biological capital' (UNEP 1995: 1). This new language is different to the language about species conservation. Genetic diversity, simultaneously both a reductive and universalising description of nature, has become an important economic phenomenon and has a value. Recent understandings of the genetic structure of all living material and new developments in the biotechnology industry are potentially raising the value of the 'biological capital' of the earth. Many scientists are now arguing that the biodiversity which has provided invaluable foods and medicines can be preserved only if there is greater monetary recognition of their value. The latest 'industrial harnessing' of life forms and processes promises new foods, medicines and energy (CIIR 1993:18). Culture is redefining the value of nature, including that of life forms themselves. The more ecocentric social movements challenge this valuing of biodiversity.

Whatever the philosophical and ethical questions raised by this conjunction of biodiversity and biotechnology, it can no longer be completely ignored or simply avoided in any contemporary conservation strategy. The pluralist position separating issues of the environment/conservation, and of development, has been severely weakened. It is now impossible for those concerned about sustainable development to talk about biodiversity without talking about biotechnology. And what are of critical importance for sustainable development are the economic and political structures within which scientific innovations are emerging. Scientific innovations in the current state of uneven development and in the current neo-liberal economic climate (see Chapter 1) could not only control nature but place it under the control of the West, at the expense of developing countries. An environmental initiative instigated by Northern conservation movements is unlikely to be able to resolve the impending economic and political restructuring of the food and pharmaceutical corporations in favour of sustainable development.

Global environmental initiatives and the burden of European colonial history

International environmental regimes are by no means new. Their main expression has been through conservation initiatives, articulated primarily through Northern social movements and NGOs since the mid-nineteenth century. Deeply embedded in the history of some Northern NGOs' history is conservation's imperial and colonial past. The North's conservation policies in the periphery started with protection of the Cape forests in order to make the wood available for use by the British Navy, and also to satisfy a newly emerging interest in tree species taxonomy. This was followed by the establishment of reserves (later to become National Parks) which again emerged as much from colonial concerns to ensure wealthy Western white males' hunting rights to big game as from growing scientific interest in the taxonomy of species. The Western interest in the taxonomy of big game was very much the outcome of identifying and making trophies out of dead species from the hunt. The hunt in Africa subsidised European expansion through the provision of meat and later the sale of ivory. Often referred to as 'The Penitent Butchers' those interested in taxonomy called various international conventions (led by the colonial powers) through the early twentieth century to encourage conservation; an early example being the establishment of the Society for the Preservation of the Wild Fauna of the Empire. Various initiatives continued through the early twentieth century and the 1949 International Union for the Conservation of Nature eventually became, in 1956, The International Union for the Protection of Nature (IUCN) which is today one of the prime movers in the conservation of biodiversity (Anderson and Grove 1987).

Likewise, the science of ecology, closely associated with Northern environmental movements, was often involved with understanding tropical environments to further colonial interests in agriculture and forestry. Many ecologists, moreover, adhere to a naturalistic model when discussing human population growth, much to the fury of the developing countries who understand their spiralling populations as the consequence of lack of development. So the South is understandably wary of any righteous conservation approach by those involved in the global arena, even when conservation stands under the umbrella of sustainable development. Development is not particularly new on the agenda of these environmentalists. What is new is the global vision of reorientating development towards Southern concerns, the notion of equity and the need for sustainability.

It was under the auspices of two longstanding and powerful Northern conservation NGOs, the IUCN and the World Wide Fund for Nature (WWF), that concern about biodiversity was addressed at Rio through UNEP. Echoing their historical positions and values, their approach was to protect and conserve biodiversity, which they have long viewed as the common heritage of humankind, by a traditional resource management approach.

The South is very wary of these NGOs, who have been instrumental in getting the environment on the recent international agenda. They claim to represent international concern for the environment and consider themselves to be above politics but the South thinks otherwise, seeing NGOs as representing Northern policy, staffed by Northern personnel and oblivious to Southern nation-states' sovereignty and their concerns about development and poverty. These conflicting attitudes are not surprising given some NGOs' long historical involvement with Southern environmental issues dating from the eighteenth and nineteenth centuries. From their inception the Northern conservation movements in the South have occupied an extremely ambivalent position. They embody the longstanding contradictions inherent in European colonial conservation policies which surface in the South's irritation with the North's hegemony in both personnel and policy. The views of many developing countries at the outset of the Convention could be summarised in the words of Shiva (1991: 7):

> While the crisis of biodiversity is focused as an exclusively tropical and Third World phenomenon, the thinking and planning of biodiversity conservation is projected as a monopoly of institutes and agencies based in and controlled by the Industrial North.

The Biodiversity Convention 1992

Both developed and developing nations avoided the divisive issues in Rio and instead agreed a Convention which simply recognised the wide-ranging implications of biodiversity conservation and use, its 'ecological, genetic, social, economic, scientific, educational, cultural, recreational and aesthetic' values; 153 heads of states signed the Convention at Rio. The United States was reluctant to sign, anxious to protect its economic interests and of course heavily lobbied by its Dedicated Biotechnology Firms and TNCs involved with the new biotechnologies. The CBD opened up new prospects for developing countires in dealing with their resources. It affirmed the sovereign rights of nation-states to their own biological resources whilst holding them responsible for conserving their biological diversity as a common concern of the international community. All usage of biological diversity was however to be in a manner compatible with sustainability. Developed countries signing the Convention agreed to contribute financially to the developing countries' conservation strategies.

The Convention sought the following commitments from national governments that signed the Convention:

> to identify and monitor biological diversity;
>
> to develop national strategies and programmes for conservation of biological diversity;
>
> to implement *in situ* (in the wild and on farm) and *ex situ* (in gene banks) conservation measures;
>
> to monitor the effects of projects on biological diversity through environmental assessment procedures;
>
> to compile national reports from all concerned parties on the effectiveness of the measures taken to implement the convention.
>
> (UNEP 1992, Convention on Biological Diversity, cited in Miller 1995: 115; Grubb and Koch 1993:76)

The North agreed to pay US$17 billion annually into a Global Environmental Fund (GEF) based within the World Bank to finance developing countries' implementation of these new conservation initiatives. Clubbe (1996) points out the insignificance of this sum compared with the US$1 trillion annual military expenditure. The South will clearly have to seek other sources of monies. In 1994 the IUCN, in order to give more substance to this global framework for the conservation of biodiversity and biological resources, proposed that national governments enact or revise their conservation legislation to provide for specified categories of protected areas in order to achieve habitat protection and hence conservation of biodiversity in the wild (*in situ* conservation). The IUCN also proposed a variety of facilities and techniques for off site (*ex situ*) conservation, including seed banks, captive breeding of animals, artificial propagation of plants for possible reintroduction into the wild and collections of living organisms in botanic gardens, zoos and aquaria for research, public education and awareness (Clubbe 1996: 221).

The value of biodiversity: a free market affair

The Convention in Article 15 affirmed the sovereign rights of nation-states to all resources including genetic resources within their national territory. Countries gained the authority to control access to these genetic resources which until now had been considered the 'common heritage' of mankind and as such freely available to all comers. All genetic resources (except genetic material held *ex situ* and collected before the Convention came into force) are from now on deemed to be of value whether wild, cultivated or improved, making biodiversity a national asset. This is particularly important for developing countries, who from now on are free to contract with interested parties not only to levy a charge on the use of resources but to agree a fair and equitable share in any results (products) of research and development of this resource 'as appropriate' and 'on mutually agreed terms'.

Biodiversity is, from now, open to monetary negotiation. The critical question is who will set the price for something which many consider to be priceless. 'The industrialised countries, corporate interests, and sectors of the scientific community' perceive the loss of common heritage 'as conflicting with their interests in having continued access to developing countries' genetic resources' (Miller 1995: 119). The developing countries will therefore be engaging in single combat on the extremely uneven battleground of the current International Economic Order (see Chapters 1 and 2). They are unprotected by any UN institution or policy or multilateral agreement despite the Convention being negotiated under the auspices of the United Nations. In effect the TNCs are likely to be highly influential in determining the price of biodiversity and the outcomes are likely to vary according to the specific strengths and weaknesses of the developing country.

The Costa Rica joint venture[1] is seen by many as the likely model for partnerships between signatory countries. Commentators are divided about the extent to which Costa Rica will gain a fair return and whether this is sufficient to ensure conservation. However, with biodiversity open to monetary gain, with the desperate needs of many developing countries for foreign exchange, and without a sophisticated multilateral agency to oversee negotiations, many future agreements might fall far short of even this agreement, particularly in relation to returning the monies to conservation and local populations. Should TNCs and national governments be left to price biodiversity through bilateral agreements? It is only the countries with a range of 'interesting' biodiversities with potential for product development that are likely to strike deals with TNCs which lead to effective local conservation initiatives. Developing countries have made it very clear they are only prepared to conserve biodiversity if it does not detract from their development priorities.

The Convention, despite on-going efforts of developing countries to improve their access to relevant technologies, has not so far held out very much promise on technology transfer. Article 16 states that access to and transfer of technology 'shall be provided and/or facilitated under fair and most favourable terms, including concessional and preferential terms'. Many commentators have pointed out that the meaning of terms like 'fair and favourable' is vague and open to interpretation, and from the developing countries' point of view do not achieve much in the way of improving terms of technology transfer which would enable them to develop appropriate new biotechnologies. Moreover, opportunities that the wording of the Convention would seem to open are closed by its concession to the current international economic order. The Convention elaborates that any technology transfer agreement between contracting parties must be 'consistent with the adequate and effective protection of Intellectual Property Rights' (cited in Grubb and Koch 1993: 79). IPRs are an extention of patent protection which protects easily copied innovations. Developing countries responding in like manner to the developed countries' resolute pursuance of IPRs want legal recognition of their rights to indigenous knowledge of their local plant and animal species. Although the Convention has clarified the position on wild genetic resources it has not attempted to negotiate cultivated or improved varieties, particularly agricultural genetic resources.

International or global governance for agrigenetic material in the International Agricultural Research Centres?

A secure global genetic resource system under global governance (in the public domain) is critical to global sustainable development. Significant biodiversity is held in national gene banks and much is *in situ* in local farmers' fields. Developing countries' participation in effective conservation programmes is critical. They are not only rich in wild biodiversity through geohistorical accident which has given them their contemporary climates and ecosystems, their vast landmasses, their prehistoric escape from periodic glaciations (ice ages), and the survival of barely inhabited primal forests. They have a wealth of social biodiversity cultivated through generations of human agricultural activity, the outcome of human history, social, cultural and scientific developments in agriculture and medicine. It is the cumulative harvest from farmers growing plants to provision themselves with foods, medicines etc. from available wild species. Social biodiversity is the outcome of protracted and intensive observation and empirical experimentation, the selection and breeding of species for locally useful genetic qualities, the matching of varieties to local soils and micro climates, and the choosing of those resistant to local pests and predators. Through effort, wisdom and design, a substantial local scientific knowledge has led to invaluable collections of genetic material in the form of new seeds and plants. This knowledge of plant diversity is maintained *in situ* by local farmers in their fields, throughout the developing world: the Mediterranean, China, India, Myanmar, South-East Asia, South-Central Asia, West Asia, Ethiopia, Middle America and South America are some of the chief centres of plant biodiversity.

The conservation of biodiversity *in situ*, on farm, is considered vital by many scientific experts working with local farmers (including women farmers) to collect, screen, conserve, improve and reintroduce indigenous seed varieties into local farming practice. Sustainable agricultural development is most likely to be progressed by a bottom-up approach which conserves biodiversity through improving and enhancing indigenous sustainable production systems in conjunction with the use of new biotechnologies. The UN Food and Agriculture Organisation (FAO) is finally acknowledging that all governments need to pay far greater attention to agricultural biodiversity through policy and financial support to local farming systems. However, international funding is not sufficiently forthcoming and the Convention did little to help these initiatives in developing countries.

Geohistoric and agriculturally derived genetic resources are also maintained *ex situ* in national and international seed banks across the world. An invaluable collection of genetic resources is stored in the International Agricultural Research Centres (IARCs).[2] These genetic resources held by the IARCs are the key to a public global genetic resource system which could further R&D into new biotechnologies for local (and global) sustainable development. The IARCs, mostly publicly funded institutions, are seen as the 'champions of the free exchange of information and technology'. Largely set up and managed by Western donors to provide research results and services to national research programmes in the Third World, they have been at the forefront of the green revolution breeding programmes to increase crop production in developing countries; for example, India, the Philippines, Mexico and Peru. They are laboratories where much of the local agricultural scientific knowledge in the form of genetic material is being conserved, stored, regenerated and made available for breeding programmes. Scientists at these centres have been involved for some time in new biotechnology programmes and are convinced of their potential for improving the productivity and quality of Third World agriculture and livestock. They are hoping to use plant genetics to improve on the inadequacies of the green revolution (Seymore 1996).

When signatories to the CBD met to agree principles to govern international trade in genetic resources a central concern was to find a way of placing the genetic resources of the IARCs under the control of all governments. However, the Convention neither clarified under whose auspices the IARCs were to continue to operate nor how their resources should be used given the increasing privatisation of the developed countries' R&D, nor what rights of access and distribution of profits should be established over the extremely valuable genetic resources held in these centres. The Convention did not see fit to set out a framework within which an assessment of the state of these resources could proceed, nor did it establish any funding mechanisms to ensure the protection of this very important collection of genetic material.

The potential value of these stocks of genetic resources is set to escalate as genes come to be routinely moved from one crop to another.[3] The fastest growing industries in Europe and especially in the United States, agribusiness and pharmaceuticals, rely on work with genes. Rio did not come up with a proposal on how to oversee the IARCs' genetic resources. In May 1994 it looked as if this genetic material would remain under global governance and in a way that represented local interests. The IARCs signed an agreement with the UN Food and Agriculture Organisation giving control to the Intergovernmental Commission on Plant Genetic Resources (ICPGR), a body sponsored by the FAO in which every country has a vote. Developing countries are in the majority on this Intergovernmental Commission and are thus ensured a prominent role in influencing decisions related to the use and value of these seed collections. Voting rights under the ICPGR could open up future possibilities of sharing in the profits that genetic engineering is likely to yield from these resources in the future.

But other actors have been trying to bring these institutes under the control of the World Bank. If the IARCs come under the control of the World Bank the developing countries will lose control to the developed world. World Bank members get votes according to their financial contributions, and the financial contributions from the developed world to the World Bank currently far outweigh those of the developing nations. Thus developed nations, via the World Bank, 'will control the shape of world agriculture for the immediate future' (Mooney cited in Mackenzie 1994). It is imperative for the development of appropriate new biotechnologies as well as for the regulation and safety of these innovative techniques that all countries can continue to access these stores of genetic diversity, and participate in decision making as to how and where new biotechnologies are used.

Unfortunately the IARCs are facing a financial crisis and might find it hard to resist the World Bank's offer to wipe out their debts and increase their funding from $7 million to $40 million a year. Recent reports of failing gene banks are putting increasing pressure on the IARCs to accept the World Bank offer. Subsequent meetings of those signing the Convention, known as the Conference of Parties (COPs), have failed so far to ensure the effective financing and overseeing at the global level of the wealth of genetic material held in the IARCs. It is critical for global sustainable development that these centres and the wealth of material they hold remain in the public sector and decision making is under global governance in order to facilitate local, diverse, agricultural biotechnologies.

The IARCs are also being adversely affected by privatisation in the developed world of R&D previously carried out in academia and public institutions. Limited funding and the growing privatisation of scientific information could undermine the ability of the IARCs to enable the development of new technologies in the public sector and to encourage their free dissemination internationally.

Novel patterns of research and development in biotechnologies replace traditional systems of open interaction and communication in research

The leading edge of biotechnology R&D is in the United States, where it increasingly takes place outside of the public laboratories of academia in the private sector. This shift into the private sector began with the rise of small Dedicated Biotechnology Firms (DBFs). These DBFs, formed around a core of talented university scientists and financed by venture capital, place contracts with academic researchers for the commercial development of laboratory discoveries. Alliances between DBFs and academia have been encouraged by the decreasing supply of government money for R&D and the limited long-term funding of academia by industry (although the US chemical firm Monsanto and the German chemical firm Hoechst have provided significant funding for academic research, the former with Washington University and the latter with Harvard).

Such collaborative ventures have been critical in financing the production and marketing of biotechnologies in the developed world, but are not available to developing countries because they cannot raise significant amounts of venture capital in their domestic markets. Avramovic (1996) suggests that with public research becoming concentrated in the larger national and international institutions, with the social North seeing a rise in integrated international research networks which by comparison have been negligible in developing countries, and with the significant rise in research in the private sector since 1992, future access for most developing countries is likely to be out of the question. Only a select few developing countries are likely to benefit. Nevertheless the governments of India and China are pushing ahead with agricultural biotechnology while Brazil and Mexico are encouraging the development of various biotechnology programmes, both national and through international collaboration.

In the current economic climate, especially where structural adjustment programmes are in force, there is limited and decreasing public funding, including the funding of public research institutes where biotechnology proceeds. Public institutes and private companies in developing countries are therefore seeking revenue from contract research and collaborating with TNCs. This means selling their services to companies abroad. Developing countries with education systems which produce skilled scientists benefit, and research is far less costly for the TNCs since salaries are lower. But TNC terms are likely to be tough and most of the biotechnology work is likely to be more relevant to foreign needs than to their domestic needs. The South in general and the IARCs institutes in particular will be adversely affected as the free flow of scientific information and germ plasm dries up. Already some IARCs are following the public institutions in the North and contracting to develop new biotechnologies with the TNCs.

A key concern is what impact will this collaboration have on developing countries? The CIP, an IARC in Peru, has entered a collaborative arrangement to develop potatoes for the fast food industry. The priority is thus to tailor for the needs of food TNCs, rather than the agronomic and nutritional requirements of farmers. Any research which results in identifying relevant genetic material that enhances potato production will be under the control of the private sector. National research programmes which should serve the interests of the majority of Third World farmers are giving way to the concerns of private industry (Hobbelink 1991:128).

The World Bank (1991) has expressed concern about this novel pattern of strategic alliances between biotechnology, academia and commerce: 'these types of contracts are rapidly altering . . . traditional systems of open interaction and communication in research'. Knowledge is increasingly falling into private hands. Private companies often obtain proprietary rights as part of the deal for developing technologies from public research and most financiers and companies are 'reluctant to become involved in biotechnology ventures unless arrangements are made to protect intellectual property rights' (cited in Avramovic 1996).

Supercombo TNCs dominate the marketing of biotechnology

Currently a few TNCs dominate the use and marketing of the new biotechnologies. Most are in the United States but TNCs from Western Europe and Japan became involved in the 1980s by making alliances and forming joint ventures with the DBFs. R&D, knowledge and technology are concentrated in a few increasingly large firms (Sharpe 1995). For example, a few supercombos dominate the production side of agribusiness; the six chemical companies dominating R&D in plant genetic engineering are Monsanto, Einmont, Du Pont (all US based), Sandoz, Zeneca and Ciba Geigy (European) (Greenpeace 1994). Increasingly large sums of money are being put into agricultural biotechnology R&D by these TNCs. In 1988 Monsanto spent US$55 million, ICI US$15 million and Du Pont US$15 million on R&D (CIIR1993: 20).

Moreover, TNCs have been furthering their corporate interests through acquisitions and mergers over the last decade or so. The most powerful ones are becoming simultaneously involved in the production of several sectors of the new biotechnologies: agrochemicals, pharmaceuticals and food. They are vertically integrating various facets of production which gives them immense control over inputs and outputs. Some of the agrochemical companies are buying up seed companies – with the intention of controlling plant breeding and crops which use their chemicals. One company, Monsanto, has as its strategic aim to define and control the framework of crop science worldwide. It has gone 'upstream', seeking control of useful genes and techniques aiming to design and sell these to breeders. It has invested huge sums in agricultural gene technology, aiming to collect the most useful gene technologies needed for the identification, transfer and patenting of useful genes. Monsanto already offers techniques for finding and introducing genes for virus resistance and herbicide tolerance and is working on resistance to fungi and improvements in quality such as better protein, oils and solids content of crops.

China seeks to protect its herbal drug treatments from exploitation by Western pharmaceutical companies

While the social North has experienced 'an explosive growth' in biotechnology in the last ten years, such R&D has been very limited in developing countries. As we have seen they are at a huge disadvantage in the race to exploit new biotechnologies and are likely to find accessing these technologies for food and pharmaceuticals extremely problematic. China is seeking to examine and develop for both the internal and export drug market its phenomenal ancient and extensive 'Materia Medica' (herbal drug treatments). It is looking for long-term investment for its own pharmaceutical industry which has seen rapid development from being worth US$190 million in 1975 to US$1.55 billion in 1992. Aware of what it stands to lose since its extensive knowledge and vast internal market could become the source of rich pickings for the TNCs, China's terms are quite tough. The Chinese authorities are insisting that any ventures are strictly for the export market and must involve the introduction of new technologies into their existing industry. China is also moving towards finding a way to accommodate Western companies' demands for more extensive patent rights without losing out on the potential of its extraordinary collection of plant biodiversity which it wishes to screen for its own development of new biotechnologies. Even if the Chinese nation benefits from these new ventures, China's medicinal herb-growing farmers stand to lose from any developments which replace their herbal medicines with drugs (Martin 1995: 6).

Northern TNCs' new biotechnologies are likely to have dramatic effects on developing countries but their products will not necessarily serve the interests of the poor. There is little interest in R&D aimed at improving many of the health problems in developing countries except where these affect the rich visitors, servicemen and tourists. R&D for developing countries' health problems has always been very much dependent on public financing and has mostly taken place in public institutions of the First and sometimes the Third World. The extreme poverty of many people and lack of government funding in the developing countries for health care inhibits private companies from embarking on costly R&D for drugs for illnesses found mostly in the social South (for example, chagas disease, malaria, leprosy) since the costs are unlikely to be recouped from the poor (see Chapter 4). In short the benefits of biopharmaceuticals are by no means equally accessible to all nations or peoples.

New biotechnologies threaten Third World trade in agricultural products

The big TNCs which have developed the new biotechnologies are also likely to have a very influential effect on global consumption and production in the area of agribusiness. Agribusiness innovations profitable for developed countries are likely to undermine Third World trade in agricultural products. In the United States genetic engineering is being developed to cut costs in the food industry through the substitution of natural products by similar genetically engineered or wholly synthetic ones. For example, US imports of sugar from the Caribbean declined by over US$400 million between 1981 and 1984, and from the Philippines by over US$600 million between 1980 and 1987, as a result of the development of genetically engineered sweeteners from maize grown in the North. If attempts to genetically engineer cocoa, palm oil and vanilla succeed then smallholders in Ghana, the Cameroon, the Ivory Coast and Zanzibar could be seriously affected. Some TNCs could be set to become universal food producers as they are moving towards selecting relevant genetic material which can be used on very basic materials to turn it into food. The total substitution of one crop for another could be a not too distant possibility and Third World farmers are likely to be severely hit by these substitution processes (CIIR 1993: 21).

Biotechnology in the hands of a few TNCs could lead to further genetic uniformity

Modern agriculture is heavily implicated in 'bio-loss . . . the world's population obtains 90% of its calories from 20 species . . . rice, maize, wheat and potatoes account for 50% of total calorie intake'. TNCs could promote extreme genetic uniformity as they engineer uniform products for mass markets worldwide. The FAO now stresses that 'intensified food production can be achieved by the sustainable use of a broader range of genetic material' but R&D doesn't necessarily coincide with the needs of the smaller farmers or with conservation interests (Panos 1995: 2). Local farmers and consumers might benefit more from genes which improve rural food crops in vulnerable and variable ecosystems, but these are less likely to make profits especially on world markets so TNCs are much less likely to develop them. The African scientists negotiating with Monsanto to use some of their new seeds made it clear that they need to identify their own technologies and that they were prepared to accept only those new genes which had been approved in the United States. They also wanted Monsanto to establish centres for breeding in Africa in order to develop the technology themselves and train local scientists in techniques of tissue culture and genetic engineering (Walgate 1990: 138–9).

Biodiversity and biotechnology: the unequal terrain of the free market

It is unlikely that biotechnology can further sustainable development within the current neo-liberal climate. The optimistic globalisation thesis of 'the level playing field' in which all participants in the world economy can benefit equally from global market forces does not appear to hold for biotechnology (see Chapter 2). The TNCs prominent at the Convention were very much in favour of the market taking care of development and the environment. The current nature and direction of global capital flows described above in relation to biotechnology suggest the new biotechnology industry does not shift that easily from one part of the world to another. It is extremely difficult for all except the largest developing countries to effectively access new biotechnologies for their own industrial developments, and relevant to their needs. The current neo-liberal approach to development is undermining longstanding public institutional structures in which new biotechnology might otherwise have been developed and disseminated to developing countries.

Patents, TRIPS and their likely impact on technology transfer and privatising genetic resources

Developed countries and the TNCs are meanwhile extremely active in other global fora to secure international agreements and policies which further their interests. Pressure from the big TNCs to increase patent protection has grown with the development of new biotechnology products. Not content with denying farmers the right to replant the new genetically engineered seeds, the TNCs are pushing through the World Trade Organisation (WTO) for biological rights as well as industrial rights. They wish to own the genetic material they have obtained where the function or application of this genetic material amounts to new knowledge. These knowledges have come to be known as Intellectual Property Rights (IPRs) or Trade-Related Aspects of Intellectual Property Rights. In essence they are seeking ownership of genetic material, i.e. lifeforms. In 1980 the US Supreme Court declared that patents could be taken out on life forms while the European Community continues to debate the legal and ethical issues surrounding proprietary rights for living material from plants and animals.

Most developing countries have weak patent laws, especially in relation to food and drugs. Many deliberately do not allow medicines or food to be patented because they are so fundamental to any society's needs. For similar reasons, developed countries did not allow patents on food, chemicals, plants or animals until the 1960s, and then only after much heated debate. The boundaries with regard to what it is both legitimate and feasible to patent have slowly shifted.[4]

Patents secure markets for a foreign company's products in other countries, especially in developing countries. By taking out a patent a TNC can prevent that country developing similar technologies, and inhibit other countries moving in with similar imports. Unless countries like India and Brazil agree to a global patent law which acknowledges TRIPS, then TNCs cannot patent their new biotechnologies in developing countries and this leaves new biotechnologies open to imitation. The developing countries have been given five years to devise patent laws (Multilateral Trade Organisation 1995). India is currently working hard to concur with this requirement by devising a patent scheme to protect their own geohistorical and social biodiversity. They are introducing new regulations which will prevent any patents being taken out on plant material which has not previously been registered under the Committee of the Indian Council of Medical Research. Terms regarding profits accruing from the plant material will be agreed on a case-by-case basis when approval is sought to take genetic material out of the country. However, national interest is not necessarily convergent with regional or local interests and the originators of some of the material, local farmers, could still go unrewarded (Jayaraman 1996). Although discussion on farmers' 'rights' is ongoing, 'Western powers have taken a cautious attitude towards this concept' (Avramovic 1996: 161).

In this context it is interesting to note that Avramovic (1996: 56) has established that US-based companies take a high proportion of patents in genetic engineering in the United States. Although this does not give an exact picture of the state of biotechnological activity by country, nevertheless, since the United States is probably the most important market in biotechnology it gives a fairly accurate picture. The United States has more than a headstart in biotechnology and the recent acknowledgement of TRIPs could prevent many developing countries from entering the race. 'If patent protection is widely extended to living matter, the existing advantages of the North in trade relations and technology will be further reinforced' (Hobbelink 1991: 115).

A 1995 Panos briefing claims that TRIPS 'represent a clear threat to world food security'. Agracetus Inc. (a major US pharmaceutical and chemical company) has already received a European patent on all transgenic soya bean varieties and with this patent goes the right to royalties from anyone else who might try to carry out biotechnology research on this crop (Panos 1995: 5). The Canadian-based Rural Advancement Foundation International (RAFI) claims Agracetus is 'working towards legal monopoly and exclusive control of soyabeans, cotton and rice'.[5] Developing countries are obviously concerned that increased patent protection law could prevent them ever using the biodiversity found within their national boundaries and undermine any attempts to develop their own biotechnologies using their knowledge derived from centuries of indigenous experimentation.

Concern about TRIPS, the demand that the Indian government introduce a Patent Law and the completion of the latest round of GATT (which enables TNCs to enforce copyright and demand royalties on any scientifically improved seeds) provoked widespread resistance in India. As the GATT talks ended there was a '10-million-strong peasants revolt. Rural farmers say their very survival is at stake.'[6]

Local farmers and indigenous people throughout the world are trying, under the auspices of the FAO, to broaden the remit of the CBD. They are currently seeking to negotiate through a Global Plan of Action, a legally binding protocol attached to the CBD which defends their interests through better crop protection and control of their crop biodiversity. Only if the Convention finds a way of acting at the local as well as the global level is it likely to secure biodiversity and appropriate biotechnologies for sustainable development.

Conclusion

The preceding analysis indicates that it is unlikely that this global vision of sustainable development can be secured through an essentially Northern environmental initiative. Conservation initiatives and their institutional frameworks are limited and fragile in the face of the development process. The ontological status and 'value' of the environment has been very explicitly made part of the global agenda of economic 'sustainable' development despite many dissenters who view this approach to nature precipitated by the development of the new biotechnologies as a crisis of modernity.

Humanity's common future, like its past, seems propelled by divisive economic and political interests. Biodiversity is not being negotiated in a political, economic or technological vacuum and is therefore not easily susceptible to global agreements. The emergence of the new biotechnologies within the current neo-liberal economic framework is currently leading to a reordering of patterns of international corporate power and trade (industrial and agricultural) in the developed and developing world. This significant global economic restructuring seems likely to further disadvantage developing countries and the poorest populations rather than to encourage sustainable development. The technocentric universalist vision of some of the scientists working for the TNCs, who hold that the new biotechnologies themselves can solve the development crisis and secure biodiversity, is unlikely to be realised.

Every nation-state clearly has enormous vested interests to protect. One of the ironies of this global convention are the implications of the process of limiting open access to local resources. The developing countries made use of the rhetoric of the global to reinforce the nation-state, territoriality and sovereignty in relation to biodiversity. Despite the Conferences of Parties at their 1994 meeting setting up a Subsidiary Body on Scientific, Technical and Technological Advice (SBSTTA) to act as a clearing house to promote and facilitate technical and scientific co-operation, focusing on the needs of developing countries, the TNCs remain in a very powerful position. It seems likely that future bilateral negotiations over sovereign states' biodiversity might only serve to further erode the collective influence of developing countries and individually they could become easy pickings in relation to the 'value' of their biodiversity. Bilateral negotiations between corporations and national governments over access, pricing and conservation of biodiversity are proving highly problematic. Post conference the difficulties facing nation-states seeking to establish ownership and claim returns on such an elusive resource are generating considerable anxiety and disillusionment.

So far, negotiating the value of biodiversity at the global level has been confined to a free market approach. Chilinski (*The Financial Times* 1996) proposed an international bank for environmental settlements which, using the environment as collateral, would match parties in environmental trade and ensure the integrity of market transactions and borrowing and lending rights. There has been no attempt to moot a global initiative to directly tax or levy TNCs' profits on biodiversity.

Unsupervised bilateral initiatives could lead to the terms of technology transfer becoming more costly and less accessible, increasing uneven development and environmental destruction. The innovative potential of the diverse new biotechnologies which Western science, with its universal applicability, in conjunction with more local *in situ* knowledges, could deliver across the planet is unfortunately unlikely to be realised in many developing countries. As the development of biotechnologies becomes increasingly confined within private institutions and conglomerates, new developments are likely to be strongly guided by the profit motive with TNCs producing agribiotechnologies with universal application. These are likely to increase uniformity even when they do develop new technologies on behalf of the interests of the poor. The current international economic system seems far from fostering sustainable patterns of trade and finance or enabling a system of agricultural production which secures the ecological base for development at the global level.

The critical need to establish the safety of these products and control their diffusion is currently being fudged through bilateral agreements between powerful conglomerates and the more vulnerable developing countries. The biotechnology industry is more closely regulated in the industrialised countries but there are few international regulations governing biosafety. Following the CBD the Conference of Parties convened a working party to develop a legally binding protocol on biosafety by 1998. The US president (Clinton) finally signed the Convention only because the US pharmaceutical and biotechnology companies became alarmed by the idea of a biosafety protocol being developed without US input (Cohen 1996). The TNCs have expressed the view that international regulations could hamper their industry.

The status of new biotechnology products in relation to corporate ownership of the biodiversity they contain is challenging international and national patent laws and aggravating many farming communities in the developing world. Following the lead of the United States, the European Union (July 1997) took the decision, despite earlier opposition from some nation-states, to patent life and to align its patent system on TRIPs within two years. It fears failure to do so would make the United States the first choice for private sector biotechnology research.

Global governance is without precedent. Unfortunately the political will to give substance to the global vision of sustainable development through the CBD is limited in the face of the free market ethos emanating from the developed world and permeating these same international institutions. Although issues are being discussed in this new global forum with the intention of protecting planet earth from the ravages of the international development process, and securing the needs of the poorest nations and people, the free market ethos and national interest continue to drive national government responses to environmental concerns. The CBD has so far, in my view, been unable to significantly influence and direct this process along a globally sustainable development path. Problems designated as global are not proving easily amenable to global solutions.

Notes

1. The agreement is between The Instituto de Biodiversidad de Costa Rica (INBio) and Merk & Co. (a US pharmaceutical conglomerate). INBio is a non-profit institution dedicated to the conservation of Costa Rican biodiversity. Its contract brief is to provide a biodiversity inventory through bioprospecting, information gathering and dissemination using paid parataxonomists – carefully trained locals – working from the park-based offices of INBio. There is also a network of local general taxonomists who liaise with international specialists. (Most specialists are found in the developed world and there are far too few of them in relation to what needs to be done to conserve biodiversity.) INBio and Merk have so far negotiated two successive, two-year bioprospecting contracts. INBio provides samples of plants, animals and micro-organisms for Merk to screen for active chemicals for use in pharmacology and for which Merk is paying $US1.3 million each contract. There are plans to expand into deals on organisms as sources of genes for biotechnology, biological control and pesticides. Ten per cent of these funds and 50 per cent of any royalties go to the Costa

Rica Ministry of Natural Resources, Energy and Mines for general conservation work, the rest is to be used for biodiversity development at INBio and in Conservation Areas (Haywood and Watson 1995 cited in Chubb 1996: 230). Whether there is any agreement on the exchange of biotechnologies is unclear.

2. Many countries have built up extensive genetic resources stored *ex situ*, in what are essentially vast refrigerators under conditions of controlled humidity and temperature. Of a total of 3.9 million seed samples held in gene banks 15.7 per cent of these *ex situ* genetic resources are held by the IARCs in 18 gene banks.

   North America holds 14.2 per cent, Europe 31.3 per cent, Latin America 14.6 per cent, Africa 6.4 per cent, and the rest of the North 7 per cent (GRAIN 1992, cited in CIIR 1993: 11). National institutes, organisations and companies with gene banks are currently starting to rent out samples to try and capitalise on these assets (*New Scientist* 21 June 1997).
3. Agribusiness is a significant earner for the developed world's economies. Australia claims to have earned $2.2 billion from IARCs' grain varieties since 1974 in increased yields; the US's billion dollar rice crop value has been increased by one-fifth and Italy's durum wheat for pasta production by $300 million. So far these profits have accrued exclusively to the countries using the genes who develop, grow and market new plant varieties. There has been no mechanism for developing countries who have contributed the genes to share in these profits. And the Rio Convention is not retrospective.
4. The legislation surrounding plants and seeds to increase protection for plant breeders was only developed in Europe through the 1960s and 1970s as plant breeding took off as an industry. In 1978 the notion of Plant Breeders Rights' (PBRs) was introduced by the International Convention for the Protection of New Varieties of Plants. PBRs granted crop breeders, mostly Northern seed companies, exclusive rights over any plant varieties they developed. These provisions downgraded farmers' rights to save seed from one harvest to plant for the next one. 'Governments are expected to continue to respect the breeders' interests as far as possible' (The Crucible Group IDRC Canada 1994). These monopolies granted under PBRs were limited to the plant variety and did not cover the genes which made up the plant.
5. W. R. Grace has also taken out US patents on four products derived from the Indian neem tree which has long been used locally in India as a medicine, a contraceptive, for toiletries and as an insecticide as well as for fuel and timber. Challenges are currently under way in the United States and Europe to revoke these patents on the grounds that these neem-based insecticides and fungicides are not innovative but derived from age-old Indian techniques (Jayaraman 1996).
6. 'In a rare episode of violence some members of the Karnataka Farmers Union ransacked the Bangalore offices of Cargill the world's largest seed company and food merchant' (*Guardian* 11 March 1994). With official bodies, National and International like the FAO, having second thoughts about the green revolution, there is a renewed and growing interest in the knowledge of local farmers.

> Local farmers are organised. They are planning to establish an international centre to back up Third World seed research and protect generations of experience and knowledge: to expand the free exchange of seeds and biological wealth

between Third World farmers and to use direct action to prevent the flow of biological wealth out of the South.'

(Searle 1994: PS)

References

Adams, W. (1990) *Green Development*, London: Routledge.

Anderson, D. and Grove, R. (1987) *Conservation in Africa, People Policies and Practices*, Cambridge: Cambridge University Press.

Anderson, J., Brook, C. and Cochrane, A. (1995) *A Global World?* Oxford: Open University Press.

Avramovic, M. (1996) *An Affordable Development*, London: Zed.

Catholic Institute for International Relations (CIIR) (1993) London: CIIR.

Chatterjee, P. and Finger, M. (1994) *The Earth Brokers*, London: Routledge.

Clubbe, C. (1996) 'Threats to biodiversity', in R. Blackmore and A. Reddish (eds) *Global Environmental Issues*, London: Hodder & Stoughton.

Cohen, B. (1996) 'Whose genes are they anyway?', *Nature* 381, 2 May.

Dobson, A. (1990) *Green Political Thought*, London: HarperCollins.

Dobson, P. (1996) *Conservation and Biodiversity*, New York: Scientific American Library.

Edwards, R. (1996) 'Seed banks fall on hard times', *New Scientist* 27, April.

Fowler, C. and Mooney, P. (1990) *The Threatened Gene*, Lutterworth Press.

Ghai, D. (1994) *Development and Environment*, Oxford: Blackwell.

Goodman, D. and Redclift, M. (1991) *Environment and Development in Latin America*, Manchester: Manchester University Press.

Grubb, M. and Koch, M. (1993) *The Earth Summit Agreement*, London: Earthscan.

Hobbelink, H. (1991) *Biotechnology and the Future of World Agriculture*, London: Zed.

Jayaraman, K. (1996) 'Gene hunters home in on India', *Nature* 381, 2 May.

Juma, C. and Mugabe, J. (1995) *Coming to Life*, Nairobi and London: ACTS Press and Zed.

Kiely, R. (1995) *Sociology and Development*, London: UCL Press.

Mackenzie, D. (1994) 'Battle for the world's seed banks', *New Scientist*, 2 July.

Martin, C. (1995) 'Over the counter', *Bulletin*, November.

Miller, M. (1995) *The Third World in Global Environmental Politics*, Buckingham: Open University Press.

Panos Media Briefing (1995) No. 17, November, London: Panos.

Pepper, D. (1993) *Eco-socialism: From Deep Ecology to Social Justice*, London: Routledge.

Perlas, N. (1994) *Overcoming Illusions about Biotechnology*, London: Zed.

Porter, G. and Brown, J. (1991) *Global Environmental Politics*, Boulder, CO: Westview Press.

Rogers, R. (1994) *Nature and the Crisis of Modernity*, Montreal: Black Rose Books.

Rosenberg, N. and Birdzell, L. (1990) 'Science, technology and the Western miracle', *Scientific American* 263 (5).

Searle, D. (1994) 'Building sustainable alternatives to GATT', *Panoscope* 38, January.

Seymore, J. (1996) 'A new revolution', *New Scientist*, 30 March.

Sharpe, M. (1995) *Biotechnology and Europe's Large Chemical/Pharmaceutical Companies*, Brighton: Science Policy Research Unit.

Shiva, V. (1991) 'Introduction', in V. Shiva and D. Cooper, *Biodiversity. Social and Ecological Perspectives*, London: Zed, pp. 1–9.

Shiva, V. and Cooper, D. (1991) *Biodiversity. Social and Ecological Perspectives*, London: Zed.

Tolba, M. *et al.* (eds) (1993) *The World Environment: 1972–1992*, UN Environment Program.

United Nations Environmental Programme (UNEP) (1995) *Global Biodiversity Assessment*, Cambridge: Cambridge University Press.

Walgate, R. (1990) *Miracle or Menace? Biotechnology and the Third World*, London: Panos Institute.

Williams, F. (1992) 'US criticism baffles backers of patents right draft', *The Financial Times*, 22 January, p. 4.

chapter 6

Globalisation, ethnic identity and popular culture in Latin America

Vivian Schelling

In recent years, sociologists, cultural critics and anthropologists have been describing, exploring, theorising and speculating on what some herald as a radically new mode of being in the world. This new mode of being, it is argued, arises out of the processes of 'globalisation', the multiple ways in which local destinies, of both nations and regions, in large parts if not most of the world, are now enmeshed with social, cultural and political forces which transcend their boundaries. According to Hannerz, we are witnessing the emergence of a world culture, which is becoming a single 'network of social relationships' characterised by a growing flow of goods, information, knowledge, images and people between different localities (Hannerz 1990).

What seems of particular significance here is that this growing interconnectedness does not principally or simply seem to entail the diffusion and imposition of Western institutions and form of culture, as theories of imperialism and Westernisation have argued. It is also not merely about the ways in which local cultures appropriate specific elements from external sources without fundamentally losing their sense of being fixed in a particular place and in the context of particular traditions.

Rather, the transformations which the term globalisation refers to are far more complex and disturbing, and this is, for several reasons. At an economic level, the activities of transnational corporations and financial institutions have acquired a global dimension accompanied by the establishment of a global financial and trading system: in cultural terms the emergence of extremely powerful culture industries (centred in the North and primarily in the United States), including television, advertising, cinema and music as well as the electronic media, compete on a global scale, creating shared desires and tastes in order to survive economically. Politically, institutions such as the United Nations, charities and non-governmental organisations operate on the basis of transnational links and transnational political cultures.

One of the most significant consequences of these processes, writers such as Giddens and Harvey have pointed out, is that social relations which were once anchored in a specific place and history have 'stretched across the globe' (McGrew 1992) in a manner which grows increasingly intense, such that 'larger and larger numbers of people live in circumstances in which disembedded institutions, linking local practices with globalised social relations, organize major aspects of day-to-day life' (Giddens 1990: 79). This 'disembedding' of cultural forms, identities, institutions and social relations, their 'scattering' across the globe and recombination in different contexts, in different space–time frames, has been defined as 'deterritorialisation', and is crucially important in understanding the cultural dimension of the 'new mode of being in the world' explored by globalisation theorists.

Deterritorialisation refers to the ways in which cultural forms and identities have migrated as a result of globalisation from their original place and reconstituted themselves in new contexts as diasporic forms. As a defence against the sense of dislocation created by deterritorialisation, this can take the shape of emphatic affirmation of the 'original' local culture or alternatively it can lead to acceptance of living in a state of 'translation' between diverse cultures. Thus the world is increasingly characterised by a proliferation of new ethnicities and fundamentalism as well as a bewildering cacophony of hybrid cultural forms and practices. What comes to mind here are the obvious examples of the new identities and practices created in particular by migrants from the 'periphery' of Europe and North America: the Asian, Caribbean, Turkish, African and Latin communities which populate the cities of the North and which either have reasserted their 'original' identity with greater vigour or developed a new more fragmented and plural sense of identity.

The speed and volatility of this process of deterritorialisation is augmented by the intervention of the culture industries, in particular music and television. Their power to transform the 'local' ethnic symbols of dislocated populations into marketable styles with a global reach, impact back onto both original and migrant cultures, transforming them and in the process creating new 'local' identifications. Thus, for example, Jean Franco points out how Latin American musicians, such as the salsa singers Celia Cruz and Ruben Blades whose music was originally rooted in specific regions, have become 'world music' superstars representing a kind of 'Latinity-within-Globalisation' (Franco 1996: 266). Via the circuits of the global media industry, South Americans both 'at home' and in the diaspora are offered an image of their specificity which is the product nevertheless of the logics of globalisation and to which each community adds its own inflection, creating a multitude of 'Latinities'. This new 'Latinity' may in some cases bear only tenuous relation to the experience – the livelihoods, sufferings, conflicts, hopes, desires, rituals and customs of those who 'originally' produced particular musical languages.

A further example illustrating this process is the effect which the export of Brazilian soap operas has had. During the 1970s, the Brazilian media conglomerate, TV Globo, gained a virtual monopoly of Brazilian television viewing time through the production of highly sophisticated soap operas, or *telenovelas*. These have become part of the staple cultural diet of the Brazilian population while also being exported to over 112 countries outside Brazil. Many of these *telenovelas* are serialisations of works of Brazilian literature, one of which in particular has had interesting 'global' effects. The *telenovela* 'A Slave called Isaura', serialisation of the novel about the trials and tribulations of a young woman born of a black slave and a white Portuguese, was shown in over eighty countries, reaching high ratings in Asia where Isaura has become a favourite name for new-born girl infants. The *telenovela* was particularly popular in China, where it was on not long after the Cultural Revolution, leading to a joint Chinese–Brazilian co-production of a *telenovela*, which in the words of *telenovela* actress Lucelia dos Santos 'will bring together two peoples and two markets' (*Folha Ilustrada* 1996).

Two important problematics arise from these new local–global formations which we will be exploring presently. One addresses what writers on Latin America refer to as the 'crisis of the popular' (Franco 1996). Due to the deterritorialisation produced by the process of globalisation, the nature and role of popular culture, the collective representations of the large mass of workers, peasants, Blacks, Indians, the un- and semi-employed has changed. Since the early twentieth century 'popular culture' in Latin America has played an important role as the repository of an independent Latin American identity or as forms of resistance – socialist, anti-imperialist, indigenous – to existing power structures. With the hybridisation of cultures produced by globalisation and the consequent blurring of boundaries between high and popular, rural and urban, oral and mass cultures, this has changed. Due to the way the global culture and tourist industries transform difference into a global 'discourse of consumerism', in other words, into 'a global currency into which all specific traditions and distinct identities can be translated' (Hall 1992: 303), it is no longer clear where the popular, either as distinctive cultural practices or as forms of resistance, is located.

The second problematic relates to the question of the role of local specificity in the context of globalisation. On the one hand, for example, Kevin Robins argues that the dynamic of global capitalist competition drives corporations such as Disney, CNN or Sky to 'target the shared habits and tastes of particular market segments at a global level, rather than marketing on the basis of geographical proximity to different national audiences' (Robins cited in Hall *et al.* 1992: 317). On the other hand, however, in order to be successful, it is also crucial to fashion products so they relate to local particularity:

> Cultural products are assembled from all over the world and turned into commodities for a new 'cosmopolitan' market-place: world music and tourism, ethnic arts, fashion and cuisine; Third World writing and cinema. The local and the exotic are torn out of place and time to be repackaged for the world bazaar. So called world culture may reflect a new valuation of difference and particularity, but it is also very much about making a profit from it.
>
> (Robins cited in Hall *et al.* 1992: 318)

Thus, differences are reinvented and reconstituted as part of the globalisation process, but as differences which appear in a commodified and marketable form, whose radical 'otherness' is tamed and domesticated. According to this view, the encounter with 'otherness' which globalisation increasingly brings about no longer necessarily entails, as it did for Europeans during the long period of colonial rule, the disquieting fear that the claims to truth and normalcy of one's own culture might be but relative and fragile constructs open to question. The 'other' here tends to become one more style or experience to taste or collect by the globally networked individual. There is however also another way of focusing on the role of the 'local' which will be relevant in understanding the forms of popular culture we will be looking at.

One of the consequences of globalisation is that, although, as Chapter 2 has pointed out, the nation-state plays an important role in the globalisation process itself, power and decision making are less clearly and firmly attached to the structures of the nation-state. It is both 'stretched', dispersed across different regions as well as highly concentrated in the hands of supranational economic and political bodies such as transnational companies, the World Bank, the

International Monetary Fund or the Group of 7. The elites of these institutions, frequently centred in the North, come together and share with local elites outside the North a common objective in managing the global economic order in their interest, forming a kind of 'transnational capitalist class', which labour movements organised more on a national basis have greater difficulty in contesting (Gill and Law 1989).

The response to this both diffuse and overwhelming constellation is manifested in the attempt by local formations at the national or regional level, in particular, social movements and NGOs, to obtain greater control of the global social forces affecting them. One expression of this can be found in the creation of local political organisations, the affirmation of local identities and creation of 'new ethnicities', not now as a style for sale in the 'world bazaar' but as movements which emanate 'from below', from 'the margins'. The growth of indigenous movements in Latin America, which in the last two decades have created transnational links between different indigenous communities in both North and South America (NACLA 1996), is one clear example of how the local and specific seeks empowerment from the margins. This is manifested not only in the organisation through which they negotiate with transnational corporations with concessions to mine or log on their land, but also in the emergence of new ethnic identities.

As Stuart Hall has observed (Hall 1991) and we shall see in our discussion of the relationship between globalisation, ethnicity and forms of popular culture in Latin America, globalisation calls forth forms of local resistance and opposition leading to what he defines as 'the margins coming into representation'. One key medium through which this occurs is by the recovery by those who exist on the margins of the prevailing global world order of their 'hidden histories':

> Face to face with a culture, an economy and a set of histories which seem to be written or inscribed elsewhere, and which are so immense, transmitted from one continent to another with such extraordinary speed, the subjects of the local, of the margin can only come into representation by, as it were, recovering their own hidden histories.
>
> (Hall 1991: 34–5)

This process of recovery of hidden histories is not a recent phenomenon but is connected to various political and intellectual transformations in the twentieth century which have questioned specific relations of domination. Historians concerned with writing 'histories from below' have produced a body of knowledge in which historical processes are analysed from the point of view of those who have had little opportunity to 'write history' due to their limited access to established educational and cultural institutions, such as workers, peasants and women. Of particular relevance for the study of globalisation and culture is the process whereby the decolonisation of the Third World has created the possibility of challenging the hegemony of Western conceptions of the individual, of reason, progress and civilisation, including the West's own conception of itself as civilisation *per se*. This process of decolonisation encompasses not only the nationalist and anti-racist movements of Africa and Asia in the post-war period but also events such as the Mexican revolution of 1910 and the Cuban revolution of 1959. These led to broader socio-economic and cultural changes in Mexico and Cuba as well as in other parts of Latin America, in which the hidden histories of slavery, the conquest and indigenous culture were partly recuperated from the silence in which they remained despite the fact that Latin America had gained its formal political independence in the nineteenth century.

Before we proceed now to investigate particular forms of popular culture, it is necessary briefly to say a few words on the relationship between 'peripheral' contexts such as Latin America and the process as well as the theory of globalisation. As David Slater (Slater 1995) has pointed out, the contemporary literature on globalisation is characterised by a limited field of vision, a 'historical and geopolitical amnesia', manifested in an apparent ignorance of the connection between contemporary global politics and the history of colonial domination, and most importantly for this study, in an apparent indifference to the 'theoretical knowledges' which are produced in the so-called 'peripheries'. It is therefore vital to rethink contemporary global theory, to 'decolonise' it, to operate an 'epistemological decentralisation' (Slater 1995: 384) through the inclusion of peripheral perspectives on the processes of globalisation, given that these remain marked by and enmeshed in the long history of unequal relations of economic and political power between the West and the non-West.

These deficiencies are all the more serious given that the symptoms noted by globalisation theorists, such as the growing hybridisation and heterogeneity of cultures, have been a characteristic of Latin America since the beginning of its history as a colonial appendage of Europe in the sixteenth century. In fact, even prior to this moment, indigenous cultures such as the Inca Empire, which imposed its culture and forms of social organisation on a vast array of conquered peoples, were of a 'hybrid' nature. Thus, it could be argued that the imposition of European culture and systems of rule on the indigenous world, the forced migration of Africans to work as slaves on colonial plantations, the highly uneven pace of development leading to the coexistence of modern and pre-modern forms of life, has made hybridity and heterogeneity constitutive of Latin America's very sense of identity. As King points out:

> because the first encounters between capitalist and pre-capitalist, between white and non-white, between the European 'Self' and the many 'Others' of the non-European world occurred in the colonies, the first globally multi-racial, multi-cultural, multi-continental societies on any substantial scale were in the periphery, not the core . . . In other words, 'Modernity' was not born in Paris but rather in Rio.
>
> (King 1991: 8)

Thus, in Latin American culture, the awareness of the 'impurity' of cultural identities, which is highlighted by globalisation theory as characteristic of the contemporary period, has always been present to a greater or lesser extent, as well as the predisposition to play with, recombine, reinvent and resignify symbols and practices connected to diverse cultural registers. This has been defined by Latin American cultural theorists using the terms 'bricolage' or 'transculturation' to denote processes through which subordinated groups selectively appropriate and transform dominant cultural influences.[1]

In considering how local popular cultures in Latin America, to which we shall now turn, are being reshaped in relation to the current context of globalisation, it is therefore important to keep in mind that earlier phases of the globalisation process going as far back as the sixteenth century had already left their mark on the local, not merely as an addition to an essentially closed and pre-formed local, but as constitutive of the identity of the local itself. Understanding the response of contemporary 'locals' therefore also entails taking account of the traces of earlier local–global relationships.

The two forms of contemporary popular culture which I will presently examine – the emergence of new expressions of black culture in Salvador, Brazil, and the development of handicraft production in Mexico – will be used to discuss and illustrate theoretical issues raised earlier with regard to the relationship between the local and the global. In particular, these two examples will be used to illuminate the distinctive ways in which expressions of the local – as the recovery of hidden histories and as a commodified style – are present and interact in complex and ambiguous ways in both these expressions of popular culture.

Salvador and the rise of a new black ethnicity

Since the mid-1970s, forms of local black popular culture have developed in the city of Salvador, Bahia, in the north-east of Brazil, which in articulation with a variety of globalising influences have become the vehicle for new expressions of black ethnicity.

The city of Salvador, the capital of Bahia, was also the capital of colonial Brazil until it was moved to Rio de Janeiro in 1763 (see map 6.1). One of the most important sources of income of the Brazilian economy in the colonial period and for most of the nineteenth century were the plantations of foodstuffs and raw materials and African slave labour on which the plantation system relied. The port of Salvador was one of the main points of entry of Africans such that Bahia, and Salvador in particular, is characterised by an extremely rich Afro-Brazilian culture with strong links to the Yoruba, Bantou and Fon origins of the black population. At present, of the 2.2 million inhabitants, 80 per cent are considered black or mestizo (Sansone 1996). However, before we focus on the new forms of local black culture which have emerged since the 1970s, it is necessary to consider the overall context of race relations in Brazil.

Race relations, and hence the position and role of black culture in Brazil, have always been most complex and full of ambiguities when compared to the Anglo-Saxon variants of race relations. As scholars comparing the two models have pointed out (Skidmore 1990), racial discrimination in Brazil is based on phenotype rather than ancestry. Moreover, in a society characterised by

Map 6.1 Brazil

widespread miscegenation and blurred racial boundaries, it is possible to achieve social mobility by whitening one's phenotype through marriage to white partners; similarly wealth and education can confer 'whiteness' on individuals. As Agier observes, racial discrimination in Brazil has worked, not through processes of rigid and clearly defined forms of exclusion and segregation, but through a peculiar form of integration in which the inequalities experienced by the black population are veiled and mitigated by intermarriage (Agier 1992).

This form of 'domination through integration' (Agier 1992) became enshrined in official ideology in the 1930s when Brazil began a process of state-led industrialisation, which required the creation of an integrated national market and the development of a distinctive national identity capable of establishing Brazil's place among the developed and 'civilised' nations. In significant ways this new national identity was characterised by the appropriation by the state and the dominant classes generally of aspects of black popular culture and their transformation into symbols of national identity. Thus the same black cultural manifestations, which in the nineteenth and early twentieth centuries were persecuted and rejected because they were seen, according to the governing Social Darwinist and Evolutionary theories of the period, as condemning Brazil to the status of a backward and uncivilised nation, now became the mark of a specifically 'tropical' and Brazilian civilisation. Gradually, samba and the Carnival celebrations of the black population of Rio de Janeiro, both cultural expressions with strong 'unruly' and oppositional elements, became practices and symbols which brought Brazilians together in a 'great communitas, where races, creeds, classes and ideologies come together peacefully to the sound of samba and racial miscegenation' (de Matta 1973: 123).

In a similar way, the Afro-Brazilian religion of Candomble, still persecuted by the authorities as a 'pagan' ritual in the 1930s, has gradually received official recognition. At the more formal ideological level, this new official attitude towards black popular culture was buttressed by the use of Gilberto Freyre's anthropological work, celebrating Brazil's legacy of miscegenation and the supposedly intimate and non-confrontational relations between blacks and whites, to elaborate what became the hegemonic discourse on race, namely, the idea that Brazil is a 'racial democracy'. This became part of the 'common sense' on race relations in Brazil to such an extent that, until fairly recently, it was seen as 'racist' to mention race relations as a possibly contentious issue. It was, as Agier points out, 'frowned upon *a priori* to talk of racism in the land of all mixtures' (Agier 1992: 61).

In this process of transformation into symbols of the 'imagined community' of modern Brazil, these popular cultural forms were partially resignified and domesticated, eliminating the more uncontrollable and unpalatable elements. The way in which these cultural processions in Rio de Janeiro have been partially transformed from celebrations of the poor into televised spectacles with considerable profits for the tourist, culture and advertising industry has been widely documented.[2]

Despite the prevalence of the ideology of racial democracy and the hegemonic absorption of aspects of black popular culture, forms of black resistance also developed simultaneously. The Frente Negra, which emerged in the 1930s, aimed at overcoming the economic and social marginalisation of blacks through education and job creation, while the MNU (Movimento Negro Unificado), a far more militant organisation created in 1978, attacked the ideology of racial democracy, denouncing the forms of racial discrimination peculiar to Brazil. In the 1980s this form of black political mobilisation was complemented and perhaps even overtaken by a new socio-cultural movement centred primarily in Salvador, in which the reinvention of black identity was linked to the processes of globalisation discussed earlier (Agier 1994).

The rise of this new form of black identity needs to be understood in the context of two further developments. The late 1970s and 1980s saw the rise of a variety of grassroots movements which were initially involved in protesting against the repressive and violent actions of the right-wing military regime. However, following the demise of the military regime in 1985, the growing power of the Labour Movement and radical Catholicism created a general climate of social and political mobilisation, an aspect of which was the defence of the rights of indigenous and black peoples and the use of ethnic symbols as a means to achieve empowerment. Simultaneously, due to Brazil's astronomical foreign debt and the consequent introduction, at the behest of the IMF, of neo-liberal economic policies, Brazil became increasingly affected by the economic and cultural dynamics of globalisation. Even though the internationalisation of the Brazilian economy was already advanced under previous governments, especially by the military government, the implementation of so-called 'structural adjustment' policies, recommended by the IMF and the World Bank, took this process further. These policies have included considerably opening up the Brazilian economy to outside influences through trade liberalisation, increased foreign investment and reduced state protection of local industry. We see thus local forces, in which the struggle for citizenship rights is connected to new ways of affirming ethnic difference and local cultures, developing at the same time as Brazilian society is opened up to powerful global economic and cultural influences.[3]

A further consideration to be taken into account is the change in the economic position of blacks in Bahia since the 1950s. With the growth of formal employment opportunities in the steel, oil and construction industries in the 1950s and 1960s, blacks gained access to a relative degree of social mobility. This entry into the formal economy, however, also led to a greater awareness of racial discrimination in that blacks as a result came face to face with the obstacles which limited their mobility (Agier 1992: 64).

It is interesting, although not surprising, that the strongholds of this new black ethnicity have been the Afro-Brazilian cultural associations connected to Carnival and Candomble. Music, dance and religious worship have been privileged sites of black cultural resistance since the colonial period. The practice of dance and music was tolerated by plantation owners and administrators as admissible forms of leisure through which, it was believed, the slave population replenished its energies. Both, however, were intimately connected with religious ritual in that they were a means of communion with African deities, which manifested themselves through states of trance induced by rhythm and song. This enabled the African population, subject to extreme forms of cultural loss and reduced to a common denominator by slavery, to reconstitute an identity in the new context of their diaspora. Following the abolition of slavery and the establishment of a republican regime in the nineteenth century, blacks were free to work as wage labourers; unskilled and marked by the stigma of slavery, they were however ill-prepared to compete on the labour market with local whites and the growing influx of Europeans migrating to Brazil in search of work. Marginalised within the new social order in which industry and urban life were becoming a significant feature, religion and music practised in the cult houses or *terreiros* of Candomble and in the black Carnival associations, both key institutions in sustaining the social networks of the urban black population, became key sources of identity and mutual support.

It is thus that in the context of the political mobilisation for citizenship rights in the 1970s and 1980s and the concomitant growth of Afro-Brazilian cultural and political groupings, the associations connected to Carnival and Candomble became particularly significant in the articulation of a new form of black ethnicity in Salvador. During this period Carnival associations with up to 20,000 members were formed, more than 2,000 Candomble temples were established, hundreds of *capoeira*[4] schools were set up, and a multitude of 'Afro' grouping of writers, dancers and musicians emerged (Agier 1992: 65).

Of all these groupings, the so-called *blocos afro* and 'afoxe groups' have been perhaps the most significant and visible expression of the development of a new black identity with up to 1,000 to 3,000 members each (Agier 1992: 66). Traditionally, the term *bloco* has been used to refer to small groups of merrymakers whose carnival disguise and dance embodies a chosen theme. 'Afoxe' groups are also defined as *blocos*, but historically they emerged out of dockworkers' efforts, who were also followers of Candomble, to publicly affirm their religion by performing and enacting aspects of religious ritual on the streets during the Carnival period. These 'afoxes' or 'street candombles' are characterised by a standard bearer with the colours, name and insignia of the group, the use of powerful percussion instruments such as the *atabaques* and *agogos*. Their songs evoke the divinities or *orishas* of the Afro-Brazilian pantheon of gods, who are also the generators of divine energy or 'axe'.

One of the first 'afoxe' groups to emerge in Salvador in the 1940s was the Filhos de Ghandi (The Children of Ghandi), connected to the dockers' trade union movements. However, in accordance with the integrationist politics pursued by the Frente Negra during this period, the Filhos de Ghandi did not pursue a black identity politics (Agier 1994). This was to occur later in the 1960s with the formation of so-called 'Indian' *blocos*, made up predominantly of blacks, and called by Brazilian Indian names, such as 'Guaranis', 'Tupis' and American Indian names such as 'Apaches' and 'Cheyennes' with which Brazilians have become familiar through North American film imports. Interestingly, these 'Indian' *blocos* used the symbols of an assumed 'other' indigenous identity – savagery and rebellion – which have long been present in debates about national identity in Brazil, as a marker of ethnic difference and as a means of expressing opposition to the heritage of slavery (Agier 1994).

In the mid-l970s, the *bloco* Ile Aye was to be one of the first to explicitly redefine local black identity, giving rise through its example to a multitude of further *blocos* and 'afoxes'. Ile Aye is a Yoruba term which initially signified 'the house of Aye', the material universe, but was then used as a term for 'black world', 'black house' (Agier 1992: 71). It was formed by a combination of civil servants, street vendors, students, dock and construction workers as a means of opening up a 'black space' within civil society. This entailed the mobilisation of specific forms of sociability and participation anchored in the collective memory of Afro-Brazilians, including religious ritual, music, dance and the engagement of the body as well as the 'rediscovery' of specific forms of African attire and Yoruba terms in order to create, as Agier suggests, a new 'ethnopolitical subject' (Agier 1992: 78). In its lyrics Ile Aye aims to recover the hidden history of slavery and the struggle for liberation, connecting contemporary struggles against racism to the memory of ancestral African kings and queens, to icons of black resistance such as Zumbi, the leader of a major slave rebellion in 1603 in the north-east of Brazil, to Nelson Mandela and Malcolm X. In the words of Joao Jorge dos Santos Rodrigues, one of the founding members of Ile Aye:

> The singers of Ile Aye have been capable of bringing together pleasure and art at the service of the community in a particular synthesis: Religion–Rebellion–Political Action–Ethnic Reeducation . . . in which dance is the concrete expression of our awareness.
>
> (dos Santos Rodrigues 1983: 248)

The reinvention and rearticulation of specific symbols have also entitled the 'reclaiming' of territory and undoing the 'deterritorialistion' experienced in the course of enslavement by, for example, redefining the neighbourhood in which Ile Aye emerged as *quilombo*, the term used for refugee slave communities.

The *afrobloco* Olodum, which has collaborated with Paul Simon and is perhaps one of the best known *blocos* outside Brazil, has also grounded its music and lyrics in the recovery of African cultural forms, and its name refers to the Yoruba term *Olodumare*, the 'god of all gods'. The followers of the *afrobloco* Muzenza, a bantou word denoting an initiate of Candomble, are inspired primarily by Rastafarianism and the work of Bob Marley (Agier 1992). This incorporation of Rastafarian symbols was taken further by other *afroblocos* through the combination of samba and reggae rhythms to form a new musical genre called samba-reggae. Similarly, the figure of the singer Gilberto Gil, frequently equated with Bob Marley, himself the icon of a Jamaican identity which was both black and modern, played a key role in the elaboration of new black ethnicity. Popularly known in the 1980s as *o poeta da raca* (the poet of the race), he was involved in defending black rights at a national level, in restoring Salvador's *terreiros* and in promulgating and internationalising this new black ethnicity through the use of rock, reggae, funk and the symbols of 'Africanness'. Moreover, Gilberto Gil's role as mediator between the local and the global was expressed in his work as cofounder, together with Caetano Veloso, of a musical genre which emerged in the 1960s and 1970s known as 'tropicalismo'. Combining bossa nova and regional musical forms with rock, rhythm and blues, jazz and experimental music, it aimed at being distinctively Brazilian while also including a global dimension.

'Tropicalismo' itself is an expression of the history and dynamics of globalisation in more than one sense. It drew on elements from the Modernist

Movement of the 1920s, in particular from one of the currents within the movement which defined itself as 'cannibalist'. Playing in an irreverent and dadaist manner with the colonial discourse which identified Brazil with savages and cannibals, the manifesto in which the 'cannibalists' stated their aims proposed that Brazil's lack of cultural independence and its subservience to Europe could be overcome through acts of 'transculturation' entailing a selective 'devouring' and indigenisation of external influences. Only thus could an autonomous local, national culture be created while making productive use of the broader global influences which since the colonial period have been constitutive of Brazilian society.[5]

Tropicalismo also conceived of itself as a response to the right-wing military coup of 1964 and the economic and political measures of the military government which accentuated the uneven development and inequalities characteristic of Brazil. With ironic allegorical images of 'Brazilian reality', dissonant collages of modern and traditional musical languages, the 'tropicalists' created a sound which embodied the jarring coexistence of widespread poverty and exploitation and the increasing internationalisation and 'modernisation' of Brazil. In that sense, 'Tropicalismo', which was centred mainly in Bahia, further developed the strategy of transculturation elaborated by the Modernists, but in order to respond to a new phase in the relationship between the local and the global.

A further factor, closely linked to global influences, which contributed to the redefinition of local black identity, is the selective appropriation by Salvadorean 'afoxe' groups of the musical and body language of soul, funk and hip-hop, which has become particularly popular in Rio de Janeiro. Introduced to Brazil, according to Vianna, without major involvement of the transitional culture industries but through exchanges between disc-jockeys of the poor quarters of Rio, this appropriation was often defended against accusations of North American cultural imperialism on the grounds that 'samba' had been coopted by the white middle classes, while 'soul' was a clear expression of the quest for black liberation (Vianna 1988).[6]

In the 1980s the municipality of Salvador conferred on Ile Aye and Olodum the title of associations of public utility. This in turn led to more concrete forms of 'reterritorialisation' in the form of official donations of land and building to the *afroblocos*. Similarly, other institutions and agents such as the Candomble *terreiros* and *capoeira* schools have extended the boundaries of this new 'ethnopolitical' identity through the formation of the Bahian Federation of Afro-Brazilian cults, the National Institute of Afro-Brazilian culture and the Council of Black Groups (Agier 1992).

This ascendancy of the diverse cultural and political groups connected to the growth of a new black ethnicity had its counterpart in the official recognition by the municipality of their growing importance and in the appropriation of this new form of 'negritude' and local Bahian identity (*baianidade*) as part of a concerted effort to promote the 'local' as a marketable style. Thus the pamphlet issued by Bahiatursa, the official Bahian Tourism Authority, defines Bahia as 'The Land of Happiness' and extols its African heritage:

> locals and foreigners dance together to the Trio Eletrico's music or to the Negro Rhythms of the Afoxes and Afro-Brazilian Carnival groups such as 'Filhos de Ghandi' and 'Olodum' . . . No wonder the locals are such happy people: the Land of Happiness is a continuous festival, usually celebrating the Catholic saints and their orisha counterparts.[7]

We can see then that the creation of a new black ethnicity and the Africanisation of *baianidade* (Bahianness) constitutes a rich and powerful affirmation of 'the local' through which a marginalised sector of Brazilian society has 'come into representation', which in turn has also improved its position within civil society. As Agier points out, the new constitution, drawn up in 1988 during the centenary of the abolition of slavery, declared that indigenous and Afro-Brazilian popular cultural forms deserved legal protection and that all forms of racial discrimination were considered illegal, thus officially disavowing the 'ideology of racial democracy'. The constitution also declared that descendants of *quilombo* (slave refugee) communities still occupying the same land had property rights to this land, thus in effect officially supporting efforts of 'reterritorialisation'. Moreover, issues concerning both the indigenous and black population have become part of party political platforms with indigenous and black people running as candidates for elections since 1980 (Agier 1994).

This 'coming into representation' then has been achieved through the recovery of a 'hidden history' entailing the elaboration of counter-narratives of Brazilian history in which being black and modern goes hand in hand with the specific appropriation of tradition. Areas of Salvador are 'reterritorialised', traditional Candomble and 'samba' groups give rise to the *afro blocos* and 'afoxes' using both Africa, past and present, black USA and Rastafarianism as 'global metaphors' through which to transcend the colonial legacy of slavery and negotiate the peculiar and slippery forms of racial discrimination characteristic of Brazil. This has nevertheless had ambiguous consequences, for, as Sansone points out, the identification with the United States, a powerful nation of the North with a tradition of black political mobilisation, has, on the one hand, enabled blacks to increase the prestige of their own subaltern ethnic culture and to demand citizenship rights and access to consumer goods. It has however also reinforced the hegemony of English-speaking diasporic cultures, reflecting the fact that globalisation is uneven and centred in the West. Moreover, the 1980s and 1990s were marked by severe economic recession in Brazil, due partly to the neo-liberal reforms introduced by the Brazilian government, leading to greater youth unemployment. This in turn has led to a transferral of status from the sphere of work to the sphere of consumption and leisure (Sansone 1996). In a globally mass-mediated society, in which style and fashion are important sources of self-esteem, this has created a tendency to use the 'deterritorialised' and somewhat uniform and stereotyped symbols of blackness, disseminated by the global culture industries, as signifiers of black identity. Paradoxically, then, the use by young Bahian blacks of US-dominated styles of blackness, such as hip-hop and black ghetto fashion, has also contributed to the growth of a local Bahian identity which circulates in the global 'world bazaar' of cultures as a deterritorialised and in many ways depoliticised 'difference'. In particular, in the brochures produced by the local authorities, a taste of 'Otherness' is being offered to global consumers in which the memory of slavery has been diluted if not eliminated altogether.[8]

By way of conclusion, it is apparent then that the emergence and flourishing of a new black Bahian identity since the 1970s reveals a tension between the two mentioned earlier. On the one hand, it constitutes an expression of local resistance emerging 'from below' and is linked to the transformations taking place in Brazil as a result of the impact of globalising forces. It could thus be argued that globalisation has not led to the 'crisis of the popular' discussed earlier, neither in the sense that 'the popular' has ceased to be the expression of subaltern resistance and agency, nor in the sense that 'the popular' can no longer be used to refer to the cultural products 'made by the people'. On the other hand, as I have tried to show, the oppositional and emancipatory affirmation of local identity overlaps and is intertwined with the expression of a black identity integrated in the global 'discourse of consumerism', connected in particular to the fashion and music industries. Here the popular exists primarily in the sense of something widely consumed rather than as the worldview and practices of subaltern groups who seek to bring about change in the social order.

Handicraft production in Mexico

The production of what is generally defined as 'handicrafts', or *artesania*, in Mexico illustrates, in a parallel yet diverse manner to the Bahian case, the way in which the prominence and visibility of 'the local' is a double-edged response to the dynamics of globalisation: partly the transformation of difference into a commodity for sale on the 'world bazaar' and partly a movement of opposition and self-defence on the part of locals.

A pervasive view on the predicament of handicraft production in modern Latin America (in particular in some Marxist and in Modernisation theory approaches) has been that as capitalist development and industrialisation advances, the textiles, pottery and masks produced by indigenous peasants for their own use, as well as the social, economic and ceremonial practices connected to them, are replaced by standardised industrial goods and mass mediated aspirations and representations. Thus, according to this view, whole ways of life are in the process of virtually disappearing. This is partly true and is evident in the increasing presence in indigenous peasant communities of televisions, cars, supermarkets, mass-produced clothing and synthetic materials. It is also apparent in the professionalisation of individual artisans and in the market-oriented transformation of designs originally connected to mythical and religious rituals, resignified as they become part of different contexts such as the museum, airport shops or elite boutiques. However, as several studies have shown, the insertion of indigenous societies into the process of capitalist development is far more complex.

Canclini's seminal study of the transformation of Tarascan *artesania* in Michoacan, Mexico, from local use objects to commodities for sale on the international market, demonstrates clearly in what ways the production of handicrafts are to be seen not simply as dysfunctional obstacles eliminated in the process of capitalist development but exist – in their very 'pre-modernness' – as key elements which sustain the reproduction of a capitalist society at various levels (Canclini 1983: 66). It has, on the one hand, been promoted by the Mexican state as a source of income to supplement precarious agricultural livelihoods and hold back the tide of rural–urban migration. On the other hand, since the 1920s, following the Mexican revolution, Mexico's indigenous heritage has also been promoted by the Mexican state as a source of national identity and cohesion, which it saw as essential for the development of state-led capitalism. Thus the claim to distinctive but undifferentiated 'Indian' identity is the other, ideological face of Mexico's attempts at nation-building, one important aspect of which consisted in the protection of national industry. Paradoxically, then, the promotion of a pre-modern 'traditional' ethnicity has gone hand-in-hand with the modernisation of Mexico. As Canclini notes: 'handicrafts and industrialisation, tradition and modernity entail each other reciprocally' (Canclini 1983: 66). It is thus through the actions of the nation-state that handicrafts gradually entered the world market where they have increasingly circulated since the 1940s and more so since the 1970s, to satisfy global elite consumer demand in the industrialised world for 'traditional' goods, which seem not to have lost their non-mechanical, individualised aura.[9] In that sense, handicraft production and the 'ethnicity' to which craft objects refer, their 'difference', function as innovations which the expansion of global capitalism, and the need to create new tastes and markets in order to maintain profitability, requires. As Stephen (1993) notes in her work on Zapotec craft production in Oaxaca, the artisans have become part of an international labour force, 'segmented on the basis of their ethnicity' (Stephen 1993: 27).

This process of incorporation within the 'dynamics of globalisation' was taken further in the 1980s, when as a result of the economic crisis in Mexico, partly due to its large external debt, the market for handicrafts became increasingly internationalised, while North American exporters, importers and retailers, who operate transnationally across the Third World, displaced Mexican intermediaries in the commercialisation of crafts (Stephen 1993: 46). Given the difference in the price of craft between Oaxaca and First World department stores, profits obtained on the world market are high and retailers take advantage of changing demands for Mexican, Indian and Persian rugs by trying to contract weavers to produce whichever design is fetching the highest price on the world market, quite independently of their anchorage in a specific history or ethnicity. This reveals that 'globalisation' is also, and very importantly so, a form of unequal exchange and that the process of cultural 'deterritorialistion', in this case the 'disembedding' of specific designs from their social context, needs to be placed in the context of unequal relations of cultural and economic power.

As in the case of Bahia, the marketing of specificity and the commodified version of indigenous identity in Oaxaca also has its counterpart in the

Map 6.2 Mexico

mobilisation by local indigenous groups of their ethnicity in a manner which is often oppositional and aims to obtain greater control over their products as well as their communities and way of life.

Stephen's study of textile production in Teotilan del Valle in the province of Oaxaca (see map 6.2) presents a revealing example of the internal changes which a handicraft-producing community goes through in the process of being incorporated in the process of capitalist development, and of the uses to which ethnicity can be put by the community in the process.

Since the rise of 'indigenism' as an ideology following the Mexican revolution, the construction of Teotiteco ethnic identity as a 'united community of weavers bonded together by kinship ties' (Stephen 1993: 28) has been used by the state to foment the idea that the artisans and their work link modern Mexico to an exalted pre-Columbian heritage as well as to promote the production and commercialisation of handicrafts. In addition, due to the influx of capital into the Teotiteco community from migrant labour returning from the United States, some capital accumulation took place in Teotitian as a whole, leading to greater class and gender differentiation within the community. This was manifested in the growing division of labour between merchants and weavers, with the latter being contracted to produce handicrafts by the former. Simultaneously, greater use was made of female labour to allow for the expansion of production. In merchant households this has led to greater subordination for women because they do not have the required management skills, in particular the ability to negotiate with foreign operators, while in independent and pieceworker households, women participate more equally in the allocation of labour and finance.

Paradoxically, then, the presentation by local merchants and foreign exporters of Teotican as a 'united community of weavers bonded together by kinship ties' with, as an importer's brochure declares, 'a 2000 year old heritage' woven from the fibres of their own strong roots (Stephen 1993: 45), has its reverse counterpart in an increasingly complex process of internal division of labour where the definition of ethnicity is contested between different groups. The meanings connected to Teotican ethnic identity, as a community of weavers connected by kinship ties, are used for several purposes: indigenous merchants use claims to their ethnicity as a negotiating tool through which to obtain higher prices and to gain greater economic control over their products; weavers whose economic position is not equal to those of merchants, but to whom they are also tied by kinship relations, use claims to ethnic unity and Teotican identity to reinforce relations of reciprocity. This ensures that more money obtained by the merchant class is invested in the community, in land, food and in the festivities and rituals through which kinship relations are articulated (Stephen 1993).

Thus we see how in this case, as in the Bahian case, ethnicity is not simply a collection of fixed cultural traits but rather, as Stephen observes, 'a contested terrain', which entails

> a group of people asserting self-generated identity based on a claim to historical autonomy and perceived natural or cultural traits that are

> emphasised as a source of identity . . . usually triggered by political, economic or social reasons in relation to one or more other social groups.
>
> (Stephen 1993: 27)

In a similar vein, it is apparent if we look at the production of other forms of handicraft, that the link between 'tradition' and 'modernity', which globalisation accelerates and which is characterised by unequal relations of power, may lead to a strengthening rather than a weakening of ethnic identity.

In Ocumicho in Michoacan, Mexico, the production of devil figures made of clay became a local tradition only a few decades ago. The clay pieces, made primarily by women (Bartra 1994), are syncretic amalgamations of traditional and modern, sacred and profane imagery (Canclini 1990). Biblical snakes, trees and holy manger scenes coexist with scenes from contemporary everyday life: buses heading for the United States, telephone conversations and aeroplanes in mid-air. Since they were first produced in the 1960s, they have been marketed by national government institutions as well as exhibited in Mexico City and New York. In the short period of thirty to forty years, this form of handicraft production has become the marker of a strong sense of local and ethnic identity. This sense of ethnic identity thus emerged, as Canclini in his account of this form of handicraft production suggests, not as a result of an attitude of closure towards modernity but rather of growing involvement with it. The way this involvement, which has given rise to a strong local tradition and ethnic identity, is negotiated by the Ocumichans is expressed in the burlesque depiction of the devils who appear as irreverent spirits in scenes of the Last Supper, holding telephones, driving aeroplanes or sitting astride on buses (Canclini 1990). The devils are, as it were, witnesses of the collision of worlds produced by the interpenetration of the local and global, the strong impact of which is mastered through laughter and irony. They very eloquently express the sense of disjuncture and dislocation produced by globalisation and the fact that we are, as Tomlinson remarks (Tomlinson 1991), constantly referred to global cultural space within which it is nevertheless most difficult to locate our own personal experience, which remains necessarily local.

Conclusion

In this chapter we have considered the nature of local responses to globalisation processes and the complex ways in which the interaction of global and local elements have given rise to new forms of ethnic identity and popular culture in Brazil and Mexico. We have examined how 'local' popular cultures and ethnic identities are ambiguously placed between functioning as commodified styles within the discourse of capitalist consumerism and as cultures of resistance which empower marginalised groups.

In the course of this investigation we have also tried to draw out features of the globalisation process which are often not, or not sufficiently, developed by globalisation theorists, namely, that globalisation is, as Kiely has stressed in the Introduction, uneven, unequal and differentially experienced. Especially in the case of the artisans of Teotilan del Valle, we saw how the affirmation of tradition and ethnic identity was used as a means of gaining greater control of the economic dynamics of globalised handicraft production. This raises further questions to be explored on the relationship between globalisation and imperialism. In other words, can local affirmations of ethnicity, as in the case of Oaxaca, be effective if their demands for greater control of the production process are not connected to those of other 'locals' in a similar predicament in the global capitalist economy? To what extent does the identification with North American black popular culture by young Bahians simply recruit a new generation for a particular form of Western popular culture and to what extent does it function as a weapon through which to challenge the Brazilian variant of race relations? Questions such a these need to be explored further if the study of globalisation processes is to go beyond looking at the multiform aspects of local–global interaction and take into account the context of national and international relations of power.

Notes

1. For a discussion of 'transculturation', see the work of Rama (1982) and Pratt (1992).
2. For a discussion of this process, see Oliven (1984, 1986), Sodre (1979) and Rowe and Schelling (1991).
3. For more on the effects of the debt crisis on Brazil see Chapter l.
4. *Capoeira* is a form of physical self-defence developed by slaves and elaborated into a dance form accompanied by samba rhythms.
5. The Modernist Movement in Brazil was launched in the 1920s by a group of painters, poets, writers and musicians. It aimed at a radical renewal of 'high culture' in Brazil which was seen by them as a mediocre imitation of European culture. It hoped to achieve this renewal through the use of new experimental aesthetic techniques as well as the incorporation of popular culture. This would, they argued, anchor Brazilian 'high culture' in its own national traditions. Among the major participants of the movement were the poets and writers Manuel Bandeira, Mario and Oswald de Andrade, the painter Anita Malfatti and the composer Villa-Lobos.
6. This is substantiated by some of the data available on the distribution of wealth in Brazil: every year the United Nations calculates a Human Development Index (HDI) for all countries (*New Internationalist* July 1996). Brazil has one of the most unequal distributions of income, in that the richest 26 per cent receive twenty-six times the income of the bottom 20 per cent. In 1970 the poorest 20 per cent had 2 per cent of household income while the richest l0 per cent had 50.6 per cent (World Bank 1987: 252–3).
7. For further discussion of this, see this volume, Chapter 7.
8. See the tourist brochure: *Brazil, Bahia, Land of Happiness* produced by Bahiatursa, the Bahia Tourist Authority.
9. See, in this respect, the work of Walter Benjamin (1970), in which he discusses the

loss of the unique 'aura' emanating from a work of art once it has been mechanically reproduced. This tears a work of art out of the particular historical and experiential moment in which it was created, destroying its authenticity and hence its 'aura', which is connected to the uniqueness of this moment.

References

Agier, M. (1992) 'Ethnopolitique: racisme, status et mouvement noir à Bahia', *Cahier des Etudes Africaines*, EHSS 32(1): 1–24.

—— (1994) 'Nation, race, culture, les mouvements noirs et Indiens au Bresil', *Cahiers des Amériques Latines* 17: 107–23.

Bartra, E. (1994) *En Busca de les Diablas*, Mexico: Tava Editorial.

Benjamin, W. (1970) 'The work of art in the age of mechanical reproduction', in W. Benjamin, *Illuminations*, Glasgow: Fontana.

Canclini, N. G. (1983) *As culturas populares no capitalismo*, Brasiliense.

—— (1990) *Culturas hibridas*, Mexico: Grijalbo.

da Matta, R. (1973) 'O Carnaval como um Rito de passagem', in *Ensaios de Antropologia Estrutural*, Petropolis: Vozes.

dos Santos Rodrigues, Joao Jorge (1983) 'A Musica de Ilê Ayê e a Educação Consciente', *Estudos Afro-Asiaticos*' 8–9: 248–51.

Franco, J. (1996) 'Globalization and the crisis of the popular', in T. Salman (ed.) *The Legacy of the Disinherited, Popular Culture in Latin America: Modernity, Globalization, Hybridity and Authenticity*, Amsterdam: Cedla.

Giddens, A. (1990) *The Consequences of Modernity*, Cambridge: Polity Press.

Gill, P. and Law, D. (1989) 'Global hegemony and the structural power of capital', *International Studies Quarterly* 33(4): 475–500.

Hall, S. (1991) 'The local and the global: globalisation and ethnicity', in A. D. King (ed.) *Culture, Globalization and the World-System*, London: Macmillan.

—— (1992) 'The question of cultural identity', in S. Hall, D. Held and T. McGrew (eds) *Modernity and its Futures*, Oxford: Polity Press.

Hannerz, U. (1990) 'Cosmopolitans and locals in world culture', in M. Featherstone (ed.) *Global Culture*, London: Sage.

Harvey, D. (1989) *The Condition of Postmodernity*, Oxford: Basil Blackwell.

Ianni, O. (1992) *A Sociedade Global*, Rio de Janeiro: Civilizacao Brasileira.

King, A. (1991) 'Introduction', in A. King (ed.) *Culture, Globalisation and the World-System*, London: Macmillan.

McGrew, A. (1992) 'A global society?', in S. Hall, D. Held and T. McGrew (eds) *Modernity and its Futures*, Oxford: Polity Press.

North American Congress on Latin America (NACLA) (1996) Report on the Americas, *Gaining ground, the indigenous movement in Latin America*, 29(5).

Oliven, R. J. (1984) 'The production and consumption of popular culture in Brazil', *Studies in Latin American Popular Culture* 3: 143–51.

—— (1986) 'State and culture in Brazil', *Studies in Latin American Popular Culture* 5: 180–5.

Pratt, M. L. (1992) *Imperial Eyes: Travel Writing and Transculturation*, London, New York: Routledge.

Rama, A. (1982) *Transculturacion narrativa en America Latina*, Mexico: Siglo XXI.

Robins, K. (1992) 'Tradition and translation: national culture in its global context', in J. Corner and S. Harvey (eds) *Enterprise and Heritage: Crosscurrents of National Culture*, London: Routledge.

Rowe, W. and Schelling, V. (1991) *Memory and Modernity: Popular Culture in Latin America*, London: Verso.

Sansone, L. (1996) 'The local and the global in today's Afro-Bahia', in T. Salman (ed.) *The Legacy of the Disinherited, Popular Culture in Latin America: Modernity, Globalization, Hybridity and Authenticity*, Amsterdam: Cedla.

Skidmore, T. (1990) 'Racial ideas and social policy in Brazil, 1870–1940', in R. Graham (ed.) *The Idea of Race in Latin America*, Austin: University of Texas Press.

Slater, D. (1995) 'Challenging Western visions of the global: the geo-politics of theory and North–South relations', *The European Journal of Development Research* 7(2): 366–88.

Sodre, M. (1979) *Samba, o Dono do Corpo*, Rio de Janeiro: Codecri.

Stephen, L. (1993) 'Weaving in the fast lane: class, ethnicity and gender in Zapotec craft commercialisation', in J. Nash (ed.) *Crafts in the World Market*, New York: State University.

Tomlinson, J. (1991) *Cultural Imperialism*, London: Pinter Publishers.

Vianna, H. (1988) *0 mundo funk carioca*, Rio de Janeiro: Zahar Editor.

World Bank (1987) *World Development Report*, Oxford: Oxford University Press.

chapter 7

Globalisation and cultural imperialism: a case study of the music industry

David Hesmondhalgh

In the 1970s and early 1980s, radical critical studies of international mass communication were dominated by the concepts of 'cultural imperialism' and 'media imperialism', and by variants of these ideas, such as 'cultural dependency' and 'electronic colonialism'. The aim of many of the writers who used these terms was to understand and criticise the way that the cultures of less-developed countries were being affected by the arrival of cultural forms and technologies (especially broadcasting) associated with the West – wealthier countries, many of which had been colonial powers right up until the 1970s, and beyond (Britain, France, Germany, but also the neo-colonialist United States). Although some Marxist writers use the term in a particular technical sense (see later), imperialism literally means the 'building of empires'. The use of the term 'cultural imperialism' implied that, now that the age of direct political and economic domination by colonial powers was (supposedly) ending, a new form of international domination was beginning. This new hegemony was based on a more indirect form of power: the fostering of cultural forms which would sap the cultural strengths of the less-developed countries, and which would allow Western-based transnational corporations to dominate non-Western economies, by encouraging a desire on the part of the post-colonial peoples for Western products and lifestyles.

It's important to realise that these writers were reacting against previous approaches to the relationship between 'West' and 'non-West' in international communications, which saw new mass media as harbingers of economic and political progress for previously colonised countries. Often referred to retrospectively as the 'modernisation' approach, this view had its heyday in the 1950s and 1960s, and was propounded by leading communications scholars such as Wilbur Schramm (Lerner 1968). Of course, many national governments in 'postcolonial' countries were fully in support of such measures. One notable case was the Shah of Iran, who poured enormous amounts of money into developing Western-style broadcasting systems (Mohammadi 1995). But the 'cultural imperialism' (from now on, CI) writers saw dangers in the export of Western communications systems. They perceived such Western (or even American) systems as threats to indigenous tradition, and attempts to persuade regional peoples of the desirability of Western-style pluralism and consumerism. The CI intellectuals thus allied themselves with regional political activists, including the first generation of post-colonial, nationalist governments in Africa and the Caribbean and nascent nationalist and anti-colonial movements. This alliance of interests was intensely critical of the (often concealed and unrecognised) class divisions within American and European societies. Why, they implicitly asked of the modernisation approach to intercultural communication, would non-Western countries want to end up like *that*? This critical alliance of scholars, political activists and national governments was at its most influential during the late 1970s and early 1980s, when its ideas found expression in a series of UNESCO reports, seminars and declarations (most notably UNESCO 1980, the McBride Report).

From the early 1980s onwards, a paradigm shift occurred in the way radical writers understood international mass communications. Those working within critical social science began to react against the CI thesis. More and more, the term used to understand the relationships between the different cultures was 'globalisation'. As the Introduction to this volume made clear, the globalisation approach argues that the different parts of the world are increasingly connected, and that the nation-state is in some sense diminishing in importance as transnational 'flows' of people, ideas and information intensify. But the way the term is used in critical social science and humanities adds further levels of connotation, beyond this relatively unproblematic kernel. The implication in much of writing from the globalisation approach is that, rather than particular nation-states attempting to impose their values on others (the cultural imperialism model), this increasing interconnectedness is essentially *undirected*; globalisation is, instead, the result of long-term shifts in the nature of political and economic systems. Whereas CI writers were influenced by Marxist critiques of the connection between national governments and private corporations, many social science writers who favour the term globalisation are somewhat more liberal in their political orientation. They might well think of certain aspects of globalisation as negative, and as favouring the interests of powerful groups, but, for them, such events are best understood, in the title of one classic study of recent global changes, as the 'consequences of modernity' (Giddens 1991).

In much of the literature on CI, the dominant form of culture was understood to be television, and the major concern was with the control and flow of *information*. While issues about the power to control the dissemination of news are crucial in any society, and while television is clearly an important mass communications medium, these emphases marginalised some important matters. Television, for example, is of minor importance in sub-Saharan Africa, where radio is the more important broadcasting form. But even where 1970s radical approaches favoured the broader notion of 'cultural imperialism' rather than 'media imperialism', culture was often defined very narrowly. The politics of popular entertainment were hardly considered. It was almost as if it was taken for granted that entertainment would be derogatory to the interests of the non-Western masses. In my view, the lack of consideration of music in these studies was particularly unfortunate. Music is a highly promising focus for case studies of international cultural and communication flows for a number of reasons. First, it makes particularly powerful statements about cultural identity, perhaps because the shared experience of rhythms and melodies helps to bind people together in particularly intense physical, emotional and sensuous ways. Second, music is highly mobile: musical styles can travel as easily as their performers, and in the modern world recorded-music commodities are highly transportable too. Cassettes, records and CDs circulate the globe, in the luggage of consumers, via television and radio broadcasts and in export crates. Third, music has the potential to cross language boundaries, because in principle its core is not linguistic but patterns of humanly organised sound. Yet paradoxically, English has become the language of global pop.

For these reasons alone, music might be a suitable basis for a reassessment of debates on CI and globalisation. But there is an additional reason. From the mid-1980s onwards, some of the most effective critiques of the CI thesis were carried out by popular music scholars (Laing 1986; Goodwin and Gore 1990; Frith 1991; Garofalo 1993). This reflected a widespread sense within popular music studies and amongst popular music fans that the popular music of the United States has actually played a progressive social role during the post-war era. Although popular music is an entertainment form which is often dominated by Western artists and companies, and can at times be mind-bogglingly trivial and banal, it has often served as a way of encouraging people to question dominant forms of power in the societies in which they live. The internationalisation of rock and roll in the 1960s is often given as an example of this. As Laing puts it:

> while rock 'n' roll was undoubtedly a moment in the expansion and technological development of the entertainment industry, it was also an instance of the use of foreign music by a generation as a means to distance themselves from a parental 'national' culture.
>
> (1986: 338)

Whether or not such relatively optimistic formulations of the political power of popular music are warranted is an issue which we shall return to. The main points I am making here are that the effectiveness of the criticisms of the CI thesis put forward by popular music writers is rooted in this sense of pop's odd radicalism; and that, again, this makes it a particularly important medium for discussion in any reconsideration of CI and globalisation.[1]

Cultural imperialism: the main lines of attack

In this chapter, I want to go against the grain of recent studies to argue that there may be some aspects of the CI thesis which it would be useful to retain; and I want to concentrate on combating the particularly strong attacks on the CI thesis which have been mounted by popular music studies. I'll begin, though, by outlining the major lines of attack on the approach in recent writing.

First, the major criticism made by those who adhere to the globalisation approach is that the CI model is *outdated*, in that the notion of imperialism, in its cultural form, or in its political and economic manifestations, relies on an analysis of the relationships between nation-states, whereas we live in an increasingly globalised world (Smith 1990). As the Introduction and Chapters 2 and 3 of this volume have pointed out, however, the nation-state is still a vital player on the international scene (see also Srerberny-Mohammadi 1996). The local and the global are highly significant, but so too is the nation: it is too early to abandon it altogether as a means of understanding modern social life. So perhaps a more convincing version of this criticism of the CI thesis is that it relies in some of its versions on a concept of national cultural autonomy which is highly problematic (Schlesinger 1987) and even at times reactionary (Mattelart *et al.* 1984).

Second, within cultural studies, the CI thesis has been accused of paying insufficient attention to *audience reception*, and the possibilities for audiences of negotiating the meanings of a text. This was the main thrust of an important early critique of the thesis by Fejes (1981); and, as more and more attention was paid to audiences in communication and cultural studies in the 1980s, this idea gained ground.

Third, the CI thesis has been acccused of being 'under-theorised'. In other words, the term has been used loosely to protest at international inequality, without thinking thoroughly about whether 'imperialism' is a justifiable term. This criticism has come from both Marxists, who point out that the term has a very specific sense in Marxist theory, as a particular stage in capitalist development (see Weeks 1991), and anti-Marxist liberal commentators, who point out that the term confuses a number of different levels of analysis. Tomlinson (1991), for example, separates out a number of different analyses contained under the banner, including criticisms of nationalism, modernity and capitalist culture.

Fourth, the CI thesis was often thought to rely on a crude conception of the effects of popular entertainment, casting America as the evil Other. Against this, it has been argued that American cultural imports could be a source of notions of rebellion and resistance in cultures where various forms of authoritarianism hold sway (for example, Reeves 1993: 207). Clearly, this relates closely to the view I referred to above, held by many popular music studies scholars, that US popular music has been the basis of progressive cultural changes.

Lastly, and related to this, the CI approach often relied on a view that sees the products of local cultures as inherently superior to internationally distributed cultural forms, on the dubious grounds that these forms spontaneously and authentically express these local cultures.

I do not have the space here to deal fully with these important issues, nor to outline the complex ways in which these criticisms are inter-related. My focus will be mainly on the position of these criticisms in discussions of popular music, and on some problems in approaches which are replacing CI methods as the principal critical perspectives available for understanding international music production and consumption. In particular, I want to look at three areas of analysis where the musical version of the thesis has come under attack.

In the first section (pp. 168–70), I examine criticisms that the 'purism' of the CI approach fails to see the value of musical *syncretism* (that is, the mixing of different musical styles, associated with different cultures). In the second section (pp. 170–6), I turn to issues of production, especially the view that shifts in the structure of the international music industry mean that we can no longer talk of CI. Here I focus on two issues: the predicted end of the dominance of the Anglo-American repertoire on the international market; and the rise of transnational corporations, not rooted in any one nation-state, and thus supposedly not representing any one type of culture. Lastly, in the third section (pp. 176–9), I probe the view that the products of the international cultural industries (exemplified here by the recording industry) are open to appropriation and recombination by local audiences. While this is obviously the case, I want to suggest some limits to how far this argument can be taken in undermining the CI thesis. My approach to these issues will be conceptual, rather than empirical. My argument is that although they may provide a basis for a considerably more subtle view of what happens at the 'local' level than in some CI writing, I believe there are important absences and problems in such revisionist work.[2]

Musical syncretism and transculturation

The most common way in which the language of CI has been rejected in debates about popular music has been to point out the aesthetic value of hybridity and syncretism in music. Syncretism is a term widely used in cultural anthropology to refer to the reconciliation of different systems of religious belief and custom. A generation of ethnomusicologists were influenced by notions of cultural influence similar to those underlying the CI thesis: complaints about the negative influence of Anglo-American forms (European harmonic systems, African-American rhythms, rock instrumentation) on local, indigenous musics were legion.[3] Such views drew on assumptions about the coherence of 'organic' cultures. Whether discussing the musical cultures of Africa or of rural England, many writers have felt that 'folk' musics are authentic cultural resources, which need to be protected against an influx of modern, commercialised sounds.[4] The orthodoxy in studies of popular music over the last fifteen years has shifted, however, to stress the fruitfulness of musical cross-fertilisation. Cultural studies has come to see all cultures as 'hybrids' of older forms, and the idea of a pure, uncontaminated tradition as problematic, and even dangerous.

The urban popular musics of West Africa provide some interesting case studies of such musical interaction and suggest some of the limitations of understanding global musical flows as threats to rooted, indigenous traditions. Christopher Waterman's study of juju (1990) details a number of interesting influences on the development of Nigerian popular music in the twentieth century, in terms of form and technology (whether electronic or acoustic-instrumental). One of the most important West African genres of the mid-century, palmwine music, was based on a mixture of 'indigenous' and imported instrumentation (guitars, mandolins, concertinas). It was heavily influenced by a number of foreign styles brought into the country via the Western technology of the gramophone, and by the British-owned multinational The Gramophone Company Ltd/EMI. These included the enormously popular country music singer Jimmie Rodgers but also Cuban groups and Hawaiian guitar. Waterman shows that even the way in which palmwine musicians tuned their guitars was similar to that of Hawaiian musicians. Yet these 'imported' styles, often the product not of an 'imperialist' culture but of other groups relatively marginalised in the global economy, were then subject to complex processes of reinterpretation. As the migrant workers of Lagos travelled between their new city and the villages where they had grown up, rural and urban guitar styles mixed.

The Yoruba ethnic group, for example, was noted for playing guitar in a style influenced by the way they played the more 'indigenous' lamellophone, for example. As sailors travelled between the major West African ports, guitar styles became diffused in new ways. Lagos palmwine playing styles were influenced by a method reputedly disseminated by Liberian sailors, for example. The product was a seemingly rich and vigorous musical culture. Although EMI undoubtedly profited from its international sales, as part of a longstanding process of political and economic domination, it is difficult to see the guitar and gramophone in this narrative as examples of a specifically *cultural* imperialism. Similarly, in an account of more recent developments in West Africa, John Collins points out that African-American influence has both helped and hindered colonial and Anglo-American domination; and he stresses the positive musical results in the area: 'Paradoxically, in copying these black American artists, Africans are being turned towards their own resources, which in recent years, has led to the creative explosion of numerous African "pop" styles' (1992: 189).

Such examples are by no means confined to West Africa. A paper by Martin Hatch on Indonesian popular music (1989) points to 'the richness and variety of musical experiences' (p. 47) in Java, encouraged rather than destroyed by the arrival of Western phenomena such as the cassette-recorder, American hard rock and John Denver. The reliance of the gramophone on electricity supplies limited access to recorded music in the 1950s, says Hatch. The relative cheapness of the cassette made recordings available to a much higher proportion of the population (though by no means everyone, as Hatch notes). Musicians adapted older genres which had come to be thought of as Indonesian (often, in fact, the result of long-forgotten interaction between, for example, Portuguese sailors and native traders) to Western rock instrumentation, to produce new pop hybrids. The strongest Western influence of all appears to have been the appearance of lyrics dealing with issues of national and regional importance. But this is hardly the kind of result that CI scholars were complaining about in the 1970s, as this kind of musical hybrid proliferated across the world. Ironically, such forms, based on American folk-rock, were probably closer to the tastes of leftist CI scholars than the traditional forms they were replacing.

One of the most compelling pieces of evidence that the impact of western-originated technological goods does not always have a detrimental effect on the diversity of expression within a particular cultural context comes from Peter Manuel's study of the impact of the introduction of cassette-recorders in North India, from the 1970s on (Manuel 1993). Manuel describes how the Indian recording industry, up until and including the 1970s, was marked by two features. The first was monopoly: GCI, the Gramophone Company of India, a subsidiary of the UK multinational EMI, completely dominated the market. The second was homogeneity of output: in the late 1970s, 90 per cent of recorded music was of one genre – film music – and this had dominated since the early 1940s. This was partly because of the enormous importance of film in India, which in turn reflected the weakness of print forms in a largely illiterate country, and the success of the cinema industry in coopting traditional theatre forms. Nearly all Indian films, which are a vastly important part of Indian popular culture, are musicals. The production of film music was completely dominated by seven or eight musical directors. Even vocal parts were confined to a handful of singers (the major star, Lata Mangeshkar, recorded 5,000 film songs between 1931 and 1980). Manuel argues that the film music genre is relatively uniform: there are virtually no significant regional influences, and few new innovations.

With the arrival of the cassette-recorder, the domination of film music and of GCI began to be challenged. At first, cassettes were common only amongst the middle class. In the 1970s, a modern form of Indian song called the *ghazal* became popular amongst this class. Like many recording industries in the developing countries, Indian recording companies were unable to break into profit because of pirate recordings. But the *ghazal* was a subtle genre which required a good quality of sound, and cheap but legal (rather than badly copied) cassettes began to sell amongst the middle class. GCI still controlled the form, but the mould was beginning to break. In the early 1980s there was a boom in devotional musics and regional styles issued on cassette by small independent companies. These cassettes were cheap to reproduce; the companies did not need to invest in the expensive pressing equipment required to press vinyl. The power requirements for cassette-players were simpler too, and player/recorders began to sell amongst other, poorer social classes. The explosion in genres continued as independent record companies found new markets. As Manuel points out, some of the new recorded genres were commercialist and chauvinist; but Manuel's overall conclusion is that the new technology of cassette-recording, introduced by Philips, a multinational capitalist corporation based in the West (the Netherlands), came to be used in a way that encouraged diversity, and that devolved ownership (Manuel 1993).

Manuel's account, like that of other popular music scholars and critical ethnomusicologists (for example, Guilbault 1993), suggests that the CI thesis would be wrong to be too pessimistic about the prospects for local music-making in an era when Western-originated technological products cross national borders. What is clear from such ethnographic accounts of generic diversity is that it is difficult to sustain a CI thesis as applied to music by relying on the concept of *homogenisation* (more generally, see the Introduction and Chapter 6 of this volume). There is simply too much evidence of continued musical heterogeneity around the world. Neither Western musical technologies nor musical genres are necessarily detrimental to the musical resources of a region. I believe, however, that the CI thesis might be more resistant to other, rather different, aspects of revisionist criticism; and in showing this, I want to deal with its potential contribution to a 'macro' level focus which would complement the stress on the local in some of the empirical studies of music I have mentioned.

Issues of production

To do this, I want to discuss some comments made by the leading popular music scholar, Simon Frith, on the 'postimperial' music industry (1991). Frith begins by questioning whether Anglo-American domination of the global music business, indisputable for the last thirty years, will continue for much longer. In the 1960s, American rock and roll, and British appropriations of it, were the basis of the international expansion of multinational companies based mainly in Britain and the United States. A convincing part of Frith's argument is that the days of British pre-eminence in repertoire (the 'Anglo' part of the combination) are numbered. Britain's important roles as a talent pool and as a test market for the world's youth are diminishing with the rise of the 'yuppie demographic and the corporate tie-in', says Frith (1991: 267). The recent emphasis in the British music press on the success of the rock band Oasis in the United States surely confirms Frith's idea, in that it reflects a peculiarly British anxiety about a decline in international cultural influence. Frith points out too that pan-European pop institutions such as fan magazines, and commercial music radio and television, mean that British bands increasingly rely on the closer European market, rather than on the American market, which is declining in economic significance anyway.

Frith, however, goes on to make a very different claim: that, as he puts it, Dutch, South African, Filipino and Hawaiian rap acts have 'as much chance to make it globally as the rappers from the Bronx or LA', as long as they sing in English (1991: 268–9). New technological and institutional conditions of music-making, such as the shift from an emphasis on live performance to recording as the main breakthrough method, mean that a new equality of access is now attainable, says Frith (267).[5] Crucially, the international record companies are engaged in new strategies whereby they are willing to distribute any locally produced music which conforms to a new global aesthetic: 'The corporations . . . control an information network, so that whatever sells in one country can be mass-marketed in another' (267). Frith's assumption is that nearly everyone has equal access to the digital technology which makes it as possible to produce the 'universalized sound', the 'global acoustic', in the Philippines as in Los Angeles. But we know that economic inequalities have often meant that access to new electric technologies have tended to be highly uneven in the past. Wallis and Malm (1984: 275), for example, show that 'acquiring and servicing the equipment of an electric group [leads] not only to new forms of economic dependence, but also to professionalization', and that this often involves the necessity of commercial sponsorship. While digital technologies may have led to a profusion of semi-professional do-it-yourself music-making in certain Western contexts, this depends on a level of disposable income not available to most of the world.

In addition, the international mass-marketing of product that 'breaks' in a particular territory is only one possible result of the control of distribution and financing exercised by the majors. Others include decisions *not* to distribute a product in certain countries which sells in others. This is especially so if this music product is sung in a language which isn't English, but often also if the sound isn't 'right' for the British and American staff who tend to make the decisions. Studies of the European recording industry have demonstrated some of these problems. Using the case of The Netherlands, Paul Rutten (1991) points to the continuing difficulties for non-American acts in 'making it' outside their own territories to show how difficult it is for acts from marginal markets to get distributed with any effectiveness on a global scale. Keith Negus too has argued that the 'dominance of Britain and the United States' in terms of repertoire in Europe is likely to continue for some time (1993: 300). Negus makes a careful distinction between the rapid development of Europe as a key, integrated market, and the rather slower and more uneven development of the different countries of Europe as repertoire sources. He detects two principal reasons for this tardy development in the attitudes of record industry personnel: a continued insistence on the aesthetic superiority of American and British acts; and ingrained habit, in that staff are unused to working with acts from other sources. This is manifested, says Negus (1993: 304), in a 'resulting mutual torpor' whereby British staff agree to accept recordings from another country in return for the release of records by lesser-known British-signed acts, but then put no effort into promotion. Rutten (1991) describes this process from the perspective of Dutch Artists and Repertoire staff: if, say, music by a Dutch artist sounds different from what is fashionable in the UK/US markets, Anglo-American staff will say it isn't suitable; if it sounds the same, they'll say that there are already plenty of British and American acts producing such music. These difficulties are almost certainly exacerbated for non-Western staff and musicians who want to break through in the most lucrative and prestigious global markets, Europe and North America.

Although they disagree on the extent to which Anglo-American dominance exists, or on whether it is likely to continue, Frith and Negus agree that the CI thesis is not a valid way to explain and analyse the current distribution of power in the international industry. Frith stresses the fact that the majors are no longer owned by companies based in the United States, but in Europe and Japan. Whereas in the recorded-music market of the 1970s the dominant companies were British (EMI and Decca), American (CBS, RCA and WEA/Warner) and Dutch/West German (PolyGram), now the spread of geographical ownership is wider. The five majors are now PolyGram (now controlled by the Dutch electronics company Philips); Sony Music Entertainment (Japan), which bought CBS in 1987; Warner Music Group (part of the US media giant Time-Warner); EMI (British owned, at time of writing); and BMG (part of the German Bertelsmann Media Group, which purchased RCA in 1986). Between them, they account for 72 per cent of the recorded-music market.[6] Not only are more of these companies now owned by non-Anglo-American parents, but they also follow different strategies from their predecessors. Frith defines these new multinationals as *distributors* of locally produced goods. He argues that we can no longer

> define the international music market in nationalistic terms, with some countries (the USA, the UK) imposing their culture on others. This does not describe the cultural consequences of the new multinationals: whose culture do Sony-CBS and BMG-RCA represent?
>
> (1991: 267)

This means, for Frith, that

> the majors don't share some supranational identity, something to be *imposed* culturally around the globe . . . The cultural imperialism model of nation versus nation must be replaced with a postimperial model of an infinite number of local experiences of (and responses to) something globally shared . . .
>
> (1991: 267–8, original emphasis)

Negus too lays emphasis on the spread of ownership and financing; this means, he says, that culture and territory can no longer be identified (1993: 311). But as Negus stresses rather more than Frith, the fact that ownership is being dispersed into a triad of East Asia, Western Europe and North America and that the multinationals are now more locally oriented than before in their strategies, does not mean that talk of cultural domination is out-dated. To be sure, it would be wrong to talk about MNCs directly representing national cultures, and was probably unwise even when most MNCs were American. This direct link of MNCs with national cultures has been widely criticised in writing on the CI thesis, and it is a just criticism. It clearly exaggerates the coherence of (at least) two very different forms of power, state and corporate-commercial.

While recognising some of these limitations in some earlier work on CI, at this point I want to raise two principal aspects of the CI literature that may be worth preserving in new approaches to the international production and circulation of music, and which have been insufficiently stressed in recent revisionist work on popular music. The first concerns the relationship between cultural goods and consumer capitalism; the second is about the continuing association of certain cultural forms with regions and nations. One significant contribution of writing under the CI banner was the argument that cultural norms, values and institutional patterns of a particular type were being foisted on to other societies (Schiller 1976; Mohammadi 1995). These included entrenched inequality of wealth, increasing concentration of power, and the weakening of civil institutions. If some other versions of the thesis tied the idea of CI too closely to the idea of nationalism and national autonomy (for example, Hamelink 1983; Nordenstreng and Schiller 1979), others have laid emphasis on this wider criticism. Criticisms of the CI thesis which (correctly) criticise its focus on the problematic concept of national autonomy leave this broader critique of capitalism untouched.

Globalisation approaches have rejected this critique of capitalism in favour of a discussion of the complex legacies of modernity. Tomlinson (1991: ch. 4) makes powerful criticisms of CI views which locate the main effects of Western 'exports' of cultural goods in their role as a harbinger or concomitant of consumer capitalism. He sees this emphasis on the role of cultural products (for example, records) in promoting ideologies such as consumerism and free market economics as a sliding towards a theory of economic domination, away from the cultural. The idea is that such Marxist-influenced theories never actually confront the moment of the cultural because of this 'economism'. For Tomlinson, cultural products are viewed by Marxian CI writers as significant only to the extent that they succeed or fail in bringing about political–economic change. Tomlinson also argues against the idea that Western or metropolitan writers can speak for the non-Western poor about what is right and wrong for them. Tomlinson is only prepared to accept the advertising of such goods as pharmaceuticals and baby products in developing countries as representative of a cultural (rather than economic) process of exploitation. He argues that the models of consumerism which CI is supposed to inculcate in the peoples of developing countries are interpreted as unhealthy only from the standpoint of metropolitan intellectuals who have experienced benefits from such models. All such talk, according to Tomlinson, is implicitly reliant on discredited Marxian notions of 'false consciousness'.

But a rejection of false consciousness approaches should not disable criticism. The agenda of international communication studies still needs to investigate the idea that the images of fame, glamour, hedonism and escapism associated with Western media products have, as one leading Latin American reviser of the CI approach puts it, qualities of 'seduction' as well as resistance (Martin-Berbero 1993). Although the kind of political mobilisation around rock music discussed in Garofalo (1992) – mass concerts for democracy in Hong Kong and Buenos Aires, for example – is important, we also need to look at some of the problems associated with the exports of such musical-ideological models: the potential to encourage apathy and passivity in the face of global commercial activity, for instance. These are issues which are difficult to treat empirically. A *lack* of political activity (and possible causes for such a lack) is inevitably harder to observe and interpret and measure than the kind of gatherings treated in Garofalo's collection. One promising avenue of study (although they make the usual gesture of accusing CI approaches of 'simplification') is indicated by Zhao and Murdock's study of the arrival of Transformer toys in China in 1989 (Zhao and Murdock 1996). The arrival of mass consumer capitalism for a quarter of the earth's population is being prepared by Western cultural brands. One of the key cultural goods in the forefront of this transformation is hi-fi (*ibid.*: 215). Although there is no question of nostalgia for feudal or Maoist China, new patterns of domination and inequality are being heralded here. The role of global Western popular-music icons will be as central as Western training-shoe brands. Central too will be regional Cantonese superstars and a new generation of Mandarin-singing stars based in Taiwan. These are home-grown stars, but their record companies mobilise distinctively Western conceptions of fame and glamour.

The second point I want to make about the globalisation critique of CI, discussed above, is that the functionalism of certain variants (for example, Schiller 1976) does not necessarily discredit the concept itself. While it is true that MNCs cannot be directly associated with a particular national culture, it is nevertheless valid to argue that they might be part of a process by which certain cultural forms, strongly associated with particular regions and nations, become predominant in terms of both sales and prestige. There is no need to attribute conscious planning or conspiracy to such companies in order to argue this. Thus the MNCs operate subject to certain economic imperatives and aesthetic histories (of the kind well described by Negus (1993) and Rutten (1991)) *with the result that* Anglo-American music is still the main type of music distributed and promoted internationally. Even if we cannot talk in functionalist terms of a conscious imposition of one culture on another, the *logic* of the global market as it is means that access to distribution and committed publicity and promotion still seem to be extremely unequal, and this inequality is geographical, and nationally differentiated. These problems can be highlighted by looking at two controversial categories, widely debated amongst fans, journalists and academics in recent years: 'Euro-pop' and 'world music'.

Continental European popular musicians have often been held in contempt by British and American audiences; 'Euro-pop' was a scathing term for the pidgin English and perceived lack of authentic musicianship amongst a breed of 1970s and 1980s acts. Laing (1992: 139) concluded a survey of national and transnational trends in European popular music by speculating that the next U2 might come from Wrocław or Bratislava. But there have been few signs of the emergence of such acts. Which continental European acts today are likely to gain the press front-cover status of U2 – or even of Meat Loaf? In the history of European acts on the global scene, only Abba have even come close to being at the centre of pop myth, and even their significance since the 1970s has been primarily based on a kitsch aesthetic. Indeed, the mid-1990s has seen the re-establishment of London as the European centre for the most fashionable pop sounds, whether in dance music or in indie/alternative pop/rock. The power centres are not shifting as quickly as some critics of CI have predicted.

The second area of debate regarding changes in international patterns of sales and prestige concerns the phenomenon of 'world music'. This was a term adapted (though hardly invented, as some writers have claimed) by a number of recording and music-press entrepreneurs to allow 'non-Western' popular musics to be promoted more adequately in Britain. Without doubt, some non-Western musicians have achieved international success and recognition. The most notable examples include Nusrat Fateh Ali Khan (Pakistan) and Youssou N'Dour (Senegal). But the impact of such musicians has been severely limited. They are enjoyed by a rather older, middle-class audience. They hardly register as popular, either in terms of total sales, or in terms of their centrality to global popular culture. In addition, such musicians are often the subject of discourses which see their music as valuable only to the extent that it conforms to certain Western notions of authenticity and tradition. Thus the notion of 'world music' as an all-embracing category for that which is not perceived as Western pop serves to exclude certain musicians from the 'centre'. What is needed then in examining the distribution of global musical resources is not only a careful, long-term consideration of patterns of sales, but also a look at which kinds of performers become part of popular-cultural mythology. For it is, arguably, at the level of fantasy and intimacy that musical culture works most powerfully. This clearly raises issues concerning the relationship of *audiences* to popular culture texts and performers. In the next section, then, I want to consider some of the possible ways of thinking about the consequences for popular music audiences of cultural inequality.

Audiences and aesthetics

Dave Laing's argument, quoted earlier (p. 166), that dominant Anglo-American musical forms can foster a creative challenge to parental national cultures (Laing 1986) is typical of a key strand of critiques of CI, the argument that it failed to pay sufficient attention to conditions of reception (see also Garofalo 1993: 25). The canonical example of such a critique in media studies is Katz and Liebes' (1985) work on *Dallas*, which aimed to show that people in different countries draw very different meanings from the programme. Such arguments clearly have important implications for how we might talk about the effects of the products of Western popular culture on the rest of the world. If those on the periphery of world power are at liberty to draw their own meanings from the products of the core, then how can we talk with any certainty about what happens as a result of their consumption of such products?

Without denying that non-Western audiences can make positive and progressive political usage from Western cultural goods, I again want to argue here for the retention of some critical lines of thought connected with the CI thesis, this time concerning what non-Western audiences can get from the music which becomes available to them, through the system of international distribution which is so thoroughly dominated by the Japanese, American and European transnationals. The first relates to the problem that much of the music which is distributed around the world is sung in English. Frith (1991: 268) argues that the very ubiquity of American culture, and the range of global appropriations of Anglo-American music, entails that to sing in English no longer means to sing in English, it 'just means to be a pop star'. He talks of the movement in Sweden to sing songs in Swedish as 'punk nationalism', a phase superseded when 'people began to realise that in today's world what matters is not just that all music from all sources should sound approximately the same, but that such sameness should also make a difference' (*ibid.*: 268).

Frith here is provocatively suggesting a new stage in the production of popular music, one where local music-making becomes aimed at international success. He offers intriguing remarks about the aesthetic dimensions of an emerging split between 'global' music and Anglo-American rock. For him, the Irish singer Sinead O'Connor represents a move towards a global music with smaller aims, 'centred on the body and the dance-floor', while her compatriots U2 are described as 'Anglo-American' in their aspirations to a global political significance (1991: 268–9). So O'Connor's music is global in its very shunning of a unifying rhetoric, its embrace of the local in the sense of small, personal; U2 are seen as affirming their (Irish) Anglo-American-ness precisely in their pursuit of grand-scale significance – their global hopes confirm them as the products of a particular local setting. Anglo-American rock is presented as reliant on a pompous rhetoric of unity. The personal, sensuous concerns of O'Connor, or the

Pet Shop Boys, are less inflated and more subtle, he implies. He suggests too that the 'old-fashioned' view that music should spread from local cult to international success is a version of the myths of rock authenticity that he has comprehensively criticised throughout his work. The 'global perspective', on the contrary, recognises the commercial and technologically mediated nature of popular music and, to a large extent, embraces it.[7]

I am sympathetic to Frith's attempts to deflate the heavy-handedness of Anglo-American rock, currently marked in many of its mainstream and 'alternative' forms by a debilitating nostalgia. It is clear that many of the most interesting recent developments in Western music are in contemporary dance music genres which reject rock aesthetics. The lack of pretentiousness in good pop music, that ability to surprise you with a sudden, seemingly unintended connection of personal feeling to a wider body of people, is indeed to be valued. But the politics of language remain submerged in Frith's account. Response to lyrics is dependent on the generic conventions at play in the particular type of music being listened to: for some genre audiences, close attention to and analysis of lyrics is vital; elsewhere, lyrics are relatively unimportant. Laing (1992) has usefully outlined a rough categorisation of the importance of language compared to vocal and musical sound in a range of genres, emphasising the different implications for different genres of the increasing integration of the European market. But while words may not be as important in pop as some middle-class audiences believe, it is still the case that some (not all) of the pleasure of music derives from the interaction of the human voice with the meaning of the words sung. No doubt it is perfectly possible for non-English-speaking audiences to enjoy, appreciate and make their own uses of a record sung in English, but we should not remain blind to the possibilities for identification opened and closed by particular language uses. Evidence that continental European audiences are turning away from global music sung in English to local musics sung in local languages may well indicate, at least in part, a desire for the pleasures of understanding.[8]

The second area of debate I want to touch on here, regarding the importance of audience identification, relates to the most striking feature of the sector of musical production adopted on a global scale: its reliance on a star system. The stars of globally successful music have, since the 1950s, nearly all come from Britain and North America. It may well be the case that, in the future, this will no longer be so. But this is a prediction, not a description. As for now, the star system provides the possibility for Anglo-American audiences of seeing local pleasures adopted on a wider, even global, scale. I want to use an anecdote to illustrate my point. I once bumped into three French students who were wandering around the part of Manchester where I used to live, and who were looking for the setting of a Smiths' song ('Rusholme Ruffians' from *Meat is Murder* (1985)). I couldn't help gaining a guilty pleasure from the incident: from the fact that someone would come all that way, from provincial France to Manchester, to try and find out more about experiences that I felt were peculiarly of the north of England, and therefore, given my own confused identifications with the region (I'm from the north-west of England), in some ways mine.

Conversely, as for many music fans, my first visit to the United States was a veritable pilgrimage of song titles and references. Although some Brits and Americans visit centres of African music, we seem to be a long way from a situation where British fans would visit Germany or Indonesia or Mongolia, in a spirit of *fandom*. These issues are not without their economic implications, given the importance of music to tourism in many localities – Liverpool, New Orleans and Kingston, Jamaica are just three examples. But these issues have a cultural importance, quite separate from their economic consequences. They help determine feelings of belonging, a sense of being marginal to systems of significance, and of being central to them. Although some writers have considered the distinction between 'core' and 'peripheral' nations in the CI literature to be outdated, as global cultural flows have become more complex, in fact it seems to me that there are still strong senses of exclusion and marginalisation at work in many places.[9]

The international distribution of industrial power, then, to some extent, structures the exclusions and inclusions that operate in the consumption of popular music. In at least some of its manifestations, the current (though perhaps increasingly vulnerable) cultural studies orthodoxy of 'active audiences' threatens to obscure this process by assuming that consumers of popular culture and popular music have equal economic and educational resources at their disposal in 'making meaning'. What I want to suggest here is that *the nature and intensity of identification allowed by popular music* might be factors we might use to assess the way the balance of power in the international recording industries operates for consumers, excluding some from certain pleasures and possibilities, and including others. There are pleasures of recognition in hearing and seeing peculiarly local ways of experiencing sexuality, violence, adolescence, class and ethnicity. These pleasures have potentially negative consequences too: they can be linked to nationalist, xenophobic exclusions of 'them' in affirming the unity of 'us'. Some of the new Hindi nationalist musics discussed by Manuel (1993) in the post-cassette proliferation of genres in Northern India fall into this category, for example. But the importance of participation on the international stage for musicians and audiences should not be underestimated. The work of Jocelyn Guilbault (1993) on *zouk*, a dance music genre particularly associated with the Lesser Antilles (especially Martinique, Dominica, Haiti and Guadeloupe), helps to make this clear, though my use of it here involves reading the grain of Guilbault's own arguments against core/periphery models. For Guilbault, *zouk* shows that the 'world music' circuits I criticised earlier can serve a positive role in allowing small and industrially developing countries to 'redefine the local', as she puts it. For her, musics such as *zouk* demonstrate that musicians from such marginalised places can make their presence felt within the market systems traditionally controlled by the dominant cultures (1993: 39). While these markets are subject to global processes of commodification, 'new alliances' are also formed with other minority groups occupying similar positions in the world and political order. Guilbault ultimately uses a typical globalisation argument to analyse the significance of *zouk* and other 'world musics' such as *rai* (Algeria) and *soukous* (Zaire): her argument is that the 'bilateral' core–periphery models characteristic of CI are too simplistic to take account of the new global interconnectedness. I have tried to show why this more traditional notion of power might still be useful; and Guilbaut's analysis helps to suggest some of the stakes involved in actually wanting to become part of the global popular-music system, rather than implicitly opting out, or 'delinking', by arguing for a notion of musical value based on separateness and indigenous tradition.

Conclusion

I have identified a number of problems with the international distribution of musical power that I think might still be usefully contained under the heading of CI. A stress on the value of musical interaction and hybridity is clearly preferable to the latent cultural purism of some thinking influenced by CI models.[10] But the CI model might still be useful in drawing our attention to important criticisms which need to be made of the capitalist production of culture as a whole: the role of musical commodities as part of wider patterns of change in leisure and entertainment, and the continued dominance of certain forms of music which, although the link between culture and territory might be growing more complex, are still linked to particular regional identities.

Musically, the logic of capitalist production has developed such that there are enormous problems of access to the world market for musicians, producers and companies who want to root their work in local experience or language, especially if these are experiences alien to the people who work for the major companies in London, New York and Los Angeles.[11] This is not to say that such local references are the only pleasures that count in popular music, or even that they are necessarily more significant politically than those afforded by the more universalised references of the singer Prince or the group U2. It is certainly not to suggest that musicians authentically express local experience through such national forms; rather, one of the fascinating features of many non-Western musics is precisely the way that such experience is mediated and fantasised, often in complex relationships with other, more global cultural forms and meanings.

I have also argued that the cultural imperialism thesis in its broadest form, as a critique of the effects of the relationships between international economic and cultural processes, is not substantially affected by the fact there has been a spread of economic power in the music industries from America and Britain to the 'global triad' of Europe, North America and South-East Asia, as long as there is no solid evidence that this spread of ownership (often hardware-driven) is substantially transforming what we might call the 'hegemonic form' of music. I argued that in terms of the cultural significance of major acts, little evidence of major change is forthcoming: very few major world stars have yet emerged from outside the Anglophone world. *If* it is fair to argue that such global music derives from European pop appropriations of African-American forms (and this is a musicological issue I have no space to argue over here) then the condoning of a global adoption of this sound risks Eurocentrism in the classic sense: the assumption and/or acceptance that specifically European/North American practices are universals.

The predominance of English as the language of pop, combined with the continued pre-eminence of Anglophone stars, some of whom still use specifically British/Irish/Australasian/North American experience as the basis of their meanings, means that industrial practice still helps to structure the possibilities of reception. If the term 'cultural imperialism' is too steeped in overtones of intention and imposition, or with associations of a past when American political and economic power coincided with cultural influence, let's be rid of it. But we shouldn't lose sight of the problems that its proponents were originally trying to draw our attention to: unequal access to the means of production, distribution, ownership, control and consumption; and its connections to a global system of consumer capitalism.

Notes

1. It is all the more striking, then, that a recent, prestigious volume devoted to reassessing the concepts of cultural imperialism and globalisation makes practically no reference to either music or to the recording industry (Golding and Harris 1997).
2. This is a heavily revised version of a paper presented at the 'European Routes' conference at the Institute of Popular Music, University of Liverpool, 21 March 1994. My thanks to Georgina Born for her comments during the early evolution of this paper. Georgie's remarks on the potential appropriateness of the CI thesis for understanding cultural developments in post-Stalinist Eastern Europe (in Born 1993) provided the initial spark for writing the paper.
3. The most notable of these ethnomusicologists was Alan Lomax (1968), whose notion of 'cultural grey-out' echoes the notion of 'cultural synchronization' in CI scholar Cees Hamelink (1983).
4. Georgina Boyes' study, *The Imagined Village* (1993), analyses the development of this attitude towards music within British society in the early part of the century, and criticises some of the problematic assumptions underlying it.
5. It is sometimes difficult to tell whether Frith is predicting or describing this new state. He begins his article by saying that he's not convinced that Anglo-American domination of popular music is 'as extensive or secure as it seems' (p. 263), suggesting a mixture of prognostication and description.
6. *Financial Times: Music and Copyright*, 31 July 1996. The figures are based on a survey of the seventeen largest national markets.
7. See also Frith (1987) for a discussion of the need for an aesthetics of popular music based around intensity of personal emotion – a discussion that addresses the issues of genre bracketed in his later (1991) article.
8. The presence of linguistic 'sub-imperialisms' such as the traditional dominance of Cantonese artists in South-East Asia should also be noted here (though these Cantonese artists are increasingly being challenged by Mandarin musicians). Chinese and Spanish-language shares of global sales are gradually increasing. As

Financial Times: Music and Copyright put it (22 November 1995: 3), 'English is no longer the only significant international language of popular music.' But it is still, for now, the most significant.

9. A significant strand of criticism of CI has been directed against the concepts of 'core' and 'periphery'. Lash and Urry (1994) provide a strong case for retaining these problematic terms, as does the Introduction to this volume – as Kiely makes clear, core and periphery cannot be equated with particular countries, given the massive inequalities within most nations. Nevertheless, it is still valid to talk of national differentiation.
10. Such interaction and cross-fertilisation have their own dangers, however: appropriation and cultural voyeurism lurk beneath the surface of some well-intentioned borrowings by western musicians (see Born and Hesmondhalgh, forthcoming); and the loss of certain musical skills and practices in developing countries cannot be merely dismissed as a sign of inevitable 'progress'. In addition, the study of, for example, generic diversity at a national level, as in Hatch's study cited above (1989), might serve to obscure long-term macro-economic and cultural processes.
11. In other words, cultural imperialism and core–periphery relations are *effects* of the workings of global cultural industries. These points are expanded on more generally in the introduction.

References

Born, Georgina (1993) 'Afterword: music policy, aesthetic and social difference', in Tony Bennett *et al.* (eds) *Rock and Popular Music: Politics, Policies and Institutions*, London: Routledge.

Born, Georgina and Hesmondhalgh, David (eds) (forthcoming) *Western Music and its 'Others': Difference, Appropriation and Representation in Music*, Berkeley: University of California Press.

Boyes, Georgina (1993) *The Imagined Village: Culture, Ideology and the English Folk Revival*, Manchester: Manchester University Press.

Collins, John (1992) 'Some anti-hegemonic aspects of African popular music', in Reebee Garofalo (ed.) *Rockin' the Boat: Mass Music and Mass Movements*, Boston, MA: South End Press.

Fejes, Fred (1981) 'Media imperialism: an assessment', *Media, Culture and Society* 3(3).

Feld, Steven (1992) 'Voices of the rainforest: imperialist nostalgia and the politics of music', *Arena* 99/100.

—— (1994) 'From schizophrenia to schismogenesis: the discourse and practices of world music and world beat', in Charles Keil and Steven Feld (eds) *Music Grooves*, Chicago: University of Chicago Press.

Frith, Simon (1987) 'Towards an aesthetic of popular music', in Richard Leppert and Susan McClary (eds) *Music and Society*, Cambridge: Cambridge University Press.

—— (1991) 'Anglo-America and its discontents', *Cultural Studies* 5(3).

Frith, Simon (ed.) (1989) *World Music, Politics and Social Change*, Manchester: Manchester University Press.

Garofalo, Reebee (1993) 'Whose world, what beat: the transnational music industry, identity, and cultural imperialism', *The World of Music* 35(2).

Garofalo, Reebee (ed.) (1992). *Rockin' the Boat: Mass Music and Mass Movements*, Boston, MA: South End Press.

Giddens, Anthony (1991) *The Consequences of Modernity*, Cambridge: Polity.

Golding, Peter and Harris, Phil (1997) *Beyond Cultural Imperialism: Globalization, Communication and the New International Order*, London: Sage.

Goodwin, Andrew and Gore, Joe (1990) 'World beat and the cultural imperialism debate', *Socialist Review* 20(3).

Guilbault, Jocelyne (1993) 'On redefining the "local" through world music', *The World of Music* 35(2).

Hamelink, Cees J. (1983) *Cultural Autonomy in Global Communication*, New York: Longmans.

Hatch, Martin (1989) 'Popular music in Indonesia', in Simon Frith (ed.) *World Music, Politics and Social Change*, Manchester: Manchester University Press.

Katz, Elihu and Liebes, Tamar (1985) 'Mutual aid in the decoding of *Dallas*', in Philip Drummond and Richard Paterson (eds) *Television in Transition*, London: British Film Institute.

Laing, Dave (1986) 'The music industry and the "cultural imperialism" thesis', *Media, Culture and Society* 8(3).

—— (1992) '"Sadeness", scorpions and single markets: national and transnational trends in European popular music', *Popular Music* 11(2).

Lash, Scott and Urry, John (1994) *Economies of Signs and Spaces*, London: Sage.

Lerner, D. (1968) *The Passing of Traditional Society*, New York: Free Press.

Lomax, Alan (1968) *Folk Song Style and Structure*, New Brunswick, NJ: Transaction.

Malm, Krister and Wallis, Roger (1992) *Media Policy and Music Activity*, London: Routledge.

Manuel, P. (1993) *Cassette Culture: Popular Music and Technology in North India*, Chicago: Chicago University Press.

Martin-Berbero, Jesus (1993) *Communication, Culture and Hegemony: From the Media to Mediations*, London: Sage.

Mattelart, Armand, Delcourt, Xavier and Mattelart, Michele (1984) *International Image Markets: In Search of an Alternative Perspective*, London: Comedia.

Mohammadi, Ali (1995) 'Cultural imperialism and cultural identity', in John Downing, Ali Mohammadi and Annabelle Srerberny-Mohammadi (eds) *Questioning the Media*, New York: Sage, 2nd edn.

Negus, Keith (1993) 'Global harmonies and local discords: transnational policies and practices in the European recording industry', *European Journal of Communication*.

Nordenstreng, Kaarle and Schiller, Herbert I. (eds) (1979) *National Sovereignty and International Communication*, Norwood, NJ: Ablex.

Reeves, Geoffrey (1993) *Communications and the 'Third World'*, London: Sage.

Robinson, Deanna Campbell, Buck, Elizabeth B., Cuthbert, Marlene and the International Communication and Youth Consortium (eds) (1991) *Music at the Margins: Popular Music and Global Cultural Diversity*, Newbury Park, CA: Sage.

Rosaldo, Renato (1989) 'Imperialist nostalgia', *Representations* 26.

Rutten, Paul (1991) 'Local popular music on the national and international markets', *Cultural Studies* 5(3).

Schiller, Herbert I. (1976) *Communication and Cultural Domination*, White Plains, NY: International Arts and Science Press.

Schlesinger, Philip (1987) 'On national identity: some conceptions and misconceptions criticized', *Social Science Information* 26(2).

Smith, Anthony D. (1990) 'Towards a global culture?', in Mike Featherstone (ed.) *Global Culture*, London: Sage.

Srerberny-Mohammadi, Annabelle (1996) 'The global and the local in international communications', in James Curran and Michael Gurevitch (eds) *Mass Media and Society*, London: Arnold, 2nd edn.

—— (1997) 'The many cultural faces of imperialism', in Peter Golding and Phil Harris (eds) *Beyond Cultural Imperialism*, London: Sage.

Straw, Will (1991) 'Systems of articulation, logics of change: communities and scenes in popular music', *Cultural Studies* 5(3).

Tomlinson, John (1991) *Cultural Imperialism: a Critical Introduction*, London: Pinter.

UNESCO (1980) *Many Voices, One World*, Paris: UNESCO.

Wallis, Roger and Malm, Krister (1984) *Big Sounds From Small Peoples*, London: Constable.

Waterman, Christopher Alan (1990) *Juju: a Social History and Ethnography of an African Popular Music*, Chicago: University of Chicago Press.

Weeks, John (1991) 'Imperialism and world market', in Tom Bottomore (ed.) *A Dictionary of Marxist Thought*, Oxford: Blackwell, 2nd edn.

Zhao, Bin and Murdock, Graham (1996) 'Young pioneers: children and the making of Chinese consumerism', *Cultural Studies* 10(2).

chapter 8

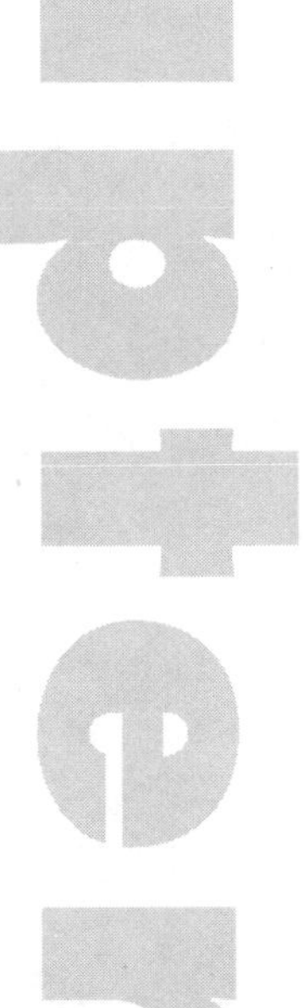

Globalisation and religious activism

Phil Marfleet

For most of the twentieth century the word 'religion' was confined to the footnotes of books dealing with the Third World. Change in Africa, Asia and Latin America – 'modernisation' – was seen as intimately connected with secularisation, with the retreat of ideas about the sacred, supernatural or otherworldly. When the term 'Third World' came into widespread use in the 1960s these assumptions meant that religion had long been viewed as merely a remnant of traditional cultures which were presumed to be undergoing linear change towards a Western model. In a comment about the Middle East which might have been made to stand for religions of the non-Western world as a whole, Daniel Lerner observed that in the face of modernisation, 'Islam is absolutely defenceless'[1] (Lerner 1964: 45). Within a decade, this assessment was thrown into question; twenty years later, with Islamic currents among the most dynamic on a world scene in which religious movements were expanding rapidly, the assertion seemed absurd.

For theorists of globalisation the increased importance of religion is consistent with socio-cultural change at a world level. Indeed, globalisation and religion are said to be intimately connected: in globalisation theory, religious resurgence is an important expression of a *unified* world. With nation-states much weakened, it is argued, supranational or transnational ideas and institutions have greatly increased in influence. As part of this process the major world religions have enjoyed an opportunity to make their world-encompassing views more relevant. The result is what Haynes calls 'a global religious revival' (Haynes 1993: 10).

Theories of globalisation have already had a profound influence on economic theory. Flows of finance, goods, information and people are said to have produced a 'borderless world' (Ohmae 1995). Differences between nation-states and regions, between 'developed' and 'developing' countries, or between 'North' and 'South', are said to be far less significant than formerly, or to have become irrelevant. 'The end of the Third World' is imminent, or even accomplished, and a new planetary order is seen to be emerging (Harris 1987, 1996). The globalist account has had a less celebrated but equally important impact on social and cultural theory. Its proponents see the emergence of a global culture in which social relations are strongly influenced by the 'singularity' of the modern world. Under these circumstances, it is argued, religious revival can be an important expression of the new 'global social reality' (Beyer 1994: 9).

This chapter looks at some problems raised by this approach, in particular by the idea that religious 'revival' is an expression of a *unifying* global order. It asks what is meant by 'the global', how religion is placed within a globalist perspective, and what this implies for an understanding of religious movements. It argues that much of the recent academic interest in religion can be traced to concern with the rise of new movements in the Third World. Here religion has seemed to assert itself most intensely, engaging tens of millions of people in movements which are not always congenial to dominant Western interests. This chapter argues that the emergence of such movements cannot be accounted for by the abstractions of globalisation theory. More appropriate, it maintains, is an approach which comprehends the concrete circumstances under which activist movements develop.

World order

Clash of civilisations

In the late 1970s, the United States government identified a novel threat to 'world order'. Following the Iranian revolution of 1979 it began to depict Islamism[2] as a force for global subversion. Muslim activists were presented as combatants in what US Secretary of State Cyrus Vance believed would be an 'Islamic–Western war' which would seriously threaten US interests (Esposito 1992: 182). At the same time, leaders of the former Soviet Union expressed their fear of 'revolutionary Islamic fundamentalism' and its threat to the Soviet state (Gupta 1986: 89). The Cold War rivals agreed that they faced a common enemy, later described by one US analyst as bent on 'global *intifada*' (Esposito 1992: 182). East and West alike viewed Lerner's 'defenceless' Islam as a major threat to their interests within the bipolar superpower arrangement and each expended great efforts in supporting governments hostile to Islamic movements.

In the 1980s, Christian radicalism was viewed similarly as a destabilising influence in the Western hemisphere. The US State Department's Council for Inter-American Security feared the growing strength of the Liberation Theology movement and especially its impact upon Latin American states closely linked to US commercial interests. It saw the Liberationists in the same light as secular radicals – as a threat to 'private ownership and productive capitalism' – and urged a 'counter-attack' against them (Smith 1991: 5). Enormous efforts were made to stem the tide of Liberationism, including attempts to direct the Roman Catholic establishment against it and to implant within the region a conservative Protestantism which might act as an ideological counterweight (Pieterse 1992; Westerlund 1996).

By the mid-1980s mass movements of religious radicalism were a feature of Third World politics, challenging the academic orthodoxy that religion must retreat in the face of modernising influences. With the exception of certain currents within North American Protestantism, all had their roots in the non-Western world. Haynes comments, 'As a result a renewed focus on the interactions between politics, theology and culture was essential in order to gain information and understanding of a Third World that stubbornly refused to conform to Western stereotypical expectations' (Haynes 1993: 2).

Some academics have sought such understanding by elaborating theories of globalisation in which religion takes on special significance. Among the most influential is the strategic analyst Samuel Huntington. His theory of world affairs suggests that states will no longer be the main players in international relations: the future will be shaped by cultural blocs defined by their religious heritage (Huntington 1993). Assuming a 'kin-country' loyalty cemented by awareness of religious heritage, Huntington argues that world events will turn on the conflicts between such blocs, producing a global 'clash of civilisations'. Heightened 'civilisation-consciousness' will emphasise ancient 'fault lines' marking territorial boundaries between the blocs. The most significant of these will be those based upon the 'Judaeo-Christian' West, the 'Confucianism/Buddhism' of East Asia and Islam, centred on the Middle East

(Huntington 1993: 23). The Islamic bloc, defined by its 'bloody borders' and still asserting itself as an 'ancient rival' of the West, will constitute the key threat to world order (Huntington 1993: 34).

This vision of an Islamic threat to the West is not novel. Huntington restates the principles of Orientalism, while formalising a notion already embraced by American strategic analysts that 'after the death of communism, Islam is the [United States'] preferred protagonist' (Esposito 1992: 5). The 'clash' theory has none the less provoked wide interest and serious debate.[3] What has attracted attention is the idea that religion can be a focal point for conflict within a *global* system. On this view the world is integrated in ways which make irrelevant old state rivalries and inequalities of wealth and power. What counts is cultural difference, most importantly the alien and threatening character of non-Western religion. In what Sakamoto calls a modern 'civilisational determinism' (Sakamoto 1995: 135), religion becomes the key category for understanding world society and 'global order'.[4]

Globalist theory

Modernity and globalisation

Interest in religion and the global condition has become so widespread that Lawrence writes of a contemporary scholarship of 'global religiosity' (Lawrence 1990: 5).[5] Notions of the 'global' which most often appear in this literature draw mainly on analyses of modernity and postmodernity produced by social and cultural theorists among whom Giddens has been particularly influential.

Giddens has argued that changes in means of communication and production have altered understandings of territorial space and chronological time. Local happenings can now be influenced almost immediately by powerful events taking place at vast distances, producing what he calls 'time–space distanciation' (Giddens 1990: ch. 4). Social relations which formerly required face-to-face contact can now operate across continents in a virtual context – time and place are in effect 'dissolved'. The result, Giddens maintains, is that social relations are 'disembedded' from their local context and that social boundaries, especially those associated with the nation-state, are broken down. Such compression of the social environment, he maintains, makes modernity 'inherently globalising' (Giddens 1990: ch. 1).

Notwithstanding that its main thrust is towards inter-relatedness, globalisation 'fragments as it coordinates', argues Giddens (Giddens 1990: 175). This point is developed by Hall, who observes that compression of space and time does not necessarily produce a more coherent social experience but generates ambivalence, uncertainty and social and cultural flux (Hall 1992). Hall and others have emphasised the question of identity within the global process, seeing individuals and groups engaged in efforts to find 'authentic', secure locations in an increasingly fluid social environment (Hall 1992; Harvey 1989;

Morley and Robins 1995). Globalisation, they argue, is not merely the imposition of a single social order or cultural practice – not merely Westernisation or the spread of consumerism – but a process that brings into being new social and cultural locations. Here, in Featherstone's words, is a 'generative frame of unity within which diversity can take place' (Featherstone 1990: 2).

For some theorists the idea of global fragmentation is more significant than the suggestion of a single social space. Identifying the emergence of a postmodernity, they stress not the relatedness of relationships, identities and images but their contingency. Axford observes that in the postmodernist account: 'Globalisation . . . appears less as the handmaiden of totalising modernisation, and more as the convenor of disorder and global restructuring' (1995: 25). In place of the attempt at theoretical integration is an emphasis on complexity and an overlapping, disjunctive order. Here, Appadurai argues, the global is best understood as a series of flows of information, images, finance, people and technologies which may conjuncturally block or facilitate the activities of power structures such as nation-states (Appadurai 1990).

Robertson and 'unicity'

For Robertson, the most influential theorist in the field of cultural globalisation, these approaches do not go far enough. He argues that today's world is not simply an enlargement or expansion of 'modern' society, or a super-postmodernity. Rather, it is a 'unicity' – a systemically united whole (Robertson 1992: 6). On this view, unifying processes have advanced so rapidly that the social environment must now be understood as 'the world as a whole' – as 'a single place' (*ibid.*: 52–3).

In this global entity, constituent parts of the system only have meaning in terms of the whole. All societies are oriented to the global or – what amounts to the same thing – to developments in other constituent parts. At the same time, the global system itself is an expression of these units' attempts to deal with their position within it. What Robertson calls 'national societies' produce histories and traditions which are their own particular account of the global order, promoting or inventing accounts of their place within this system, so that everywhere there is concern about the universal – about the general applicability of particular accounts of the global. Robertson concludes that there is a 'global universal' which emerges from the interaction of the complementary images of local societies. The net effect is to produce 'intensification of consciousness of the world as a whole' (Robertson 1992: 10).

A 'global field' has emerged, argues Robertson. This is a socio-cultural system which is the product of compression of 'civilisational structures, national societies, intra- and cross-national movements and organisations, sub-societies and ethnic groups, intra-societal groups, individuals and so on' (Robertson 1992: 61). Particular interpretations of the human condition are marked by their place in the global field, says Robertson. There is a simultaneous process under way in which universal and particular ('local') elements are both active. The result is: 'the particularisation of the universalism (the rendering of the world as a single place) and the universalisation of particularism (the globalised expectations that societies . . . should have distinct identities)' (Beyer 1994: 28). For Robertson, globalisation involves the crystallisation of suppositions about what the world is and what it should be. Under global circumstances both dominant suppositions and alternative suppositions which challenge them influence 'the trajectories of globalisation' (Robertson 1992: 68). Among such alternative perspectives, he argues, are religious currents which are prominent in 'the search for fundamentals' – the attempt to locate identity in a changing world.

The search for identity is also global, Robertson insists, in that it is pursued in terms of ideas about tradition, history, locality, community and nation which have been globally diffused. Today, religious currents must be understood as expressions of their place within the global field.

Globalisation and religion

For social and cultural theorists who have focused on religion the implications of these analyses are profound. First, it is argued, changes in means of communication have altered the ways in which religious beliefs operate to confirm and assert social identity. Introducing his *Religion and Globalization*, Peter Beyer observes:

> The power of . . . technology makes very rapid communication possible over almost unlimited space. Moreover [the new] media exist nearly everywhere on earth, along with the will and ability to use them. The potential for worldwide communication has been translated into actual practice. We therefore live in a *globalizing* social reality, one in which previously effective barriers to communication no longer exist.
>
> (Beyer 1994: 1)

The result, argues Beyer, is that people, cultures, societies and civilisations previously more or less isolated from one another are now in regular, almost unavoidable, contact (Beyer 1994). National borders, in particular, are of less and less significance. Such changes are said to have special significance for religion because as social space contracts there may be much greater awareness of the diversity of human experience, challenging local belief systems and the worldviews or 'life worlds' they encompass.[6] The modern world is also highly unstable, uncertain and demanding of interpretation. The global context is therefore also one which encourages, even insists upon, the assertion of worldviews, promoting the search for fundamentals. These may be projected or amplified within a global system free of the old barriers to communication. Localisms, with their attempts to reassert particular histories and to project particular identities, are the more able to assert their universalist dimensions.

On this view, the qualities of religions as worldviews make them especially significant within the globalising process. The transcendences and universalisms of the major world religions are complemented by the new possibilities of projecting them across regions, nations and continents. Waters observes:

> For many centuries, the great universalising religions of the world, Buddhism, Christianity, Confucianism, Islam and Hinduism, offered adherents an exclusivist and generalising set of values and allegiances that stood above both state and economy . . . These religions in particular have had a globalising sense of mission . . .
>
> (Waters 1995: 125)

Now such a mission, it is argued, may be conducted within a congenial global context.

For Beyer, globality implies completion of 'mission'. Under the new circumstances, 'the crystallisation of telic concerns' that characterises globalisation provides an appropriate medium for world theologies that address 'the eschatological implications of an inclusive humanity' (Beyer 1994: 30). The universalism of religion, it is argued, meets the global at the level of world inclusiveness. The modern order has brought intense pressure to bear upon older worldviews; at the same time it has offered them vast new potentials. A globalised world, in short, supercharges religion.[7]

Most theorists in this field accept the postmodernist account of the disruption of certainties and the critique of 'meta-' or 'master-narratives' such as the notions of progress and development. They view the search for 'fundamentals' associated with contemporary religious currents as an expression of hostility to aspects of the global system such as economic penetration, secularism and the influence of 'alien' values which *at the same time* rejects the system's perceived dislocation and moral relativism. In proposing encompassing solutions, it is said, this quest does not *challenge* globalising processes but effectively complements them. What Robertson calls (with a religious allusion) 'global callings' constitute a search for reference points which can only further globalisation. Beyer observes that, 'The intent is to shape the global reality, not to negate it' (Beyer 1994: 3).

The net effect for Beyer is that religion has become 'a proactive force'. It is now 'instrumental' in the elaboration and development of globalisation. He concludes that 'the central thrust is to make [believers] more determinative in the world system' (Beyer 1994: 3).

Religion and society

At one level global theory is uncontentious. Changes in means of communication, and in flows of resources and of people, can clearly contribute to the erosion of social boundaries and bring greater interconnectedness. Such processes have been at work for centuries.[8] But 'strong' theories of globalisation, which promote the idea of of a novel, systemic unity, present more serious problems.

A series of recent studies has questioned notions of globality in the fields of economics and political economy (Harman 1996; Hirst and Thompson 1996; Hoogvelt 1997; Kiely 1995). They have suggested that the premises of globalisation theory are unsustainable; in particular, that the idea of a world unified by capital flows conceals increasing inequality and asymmetry. Most important in the current context is the conclusion that most of Africa, Asia and Latin America will remain marginal to the world economy and largely powerless to influence world events – that the old duality of First and Third Worlds has not been modified by 'globalising' forces but has been reinforced, even exaggerated.[9]

These criticisms, and those made by Kiely elsewhere in this book (Introduction and chapter 2), place a very serious question mark over the idea of a new global paradigm. I want to examine some of the assumptions within related socio-cultural theories of globalism and consider their consequences for religion. In doing so, I want to suggest that models which underlie these theories in fact inhibit understanding of religious belief and practice, especially religious activism in the Third World. They also have the effect of falsely presenting social movements in general as consistent with an integrated world.

What is religion? Definitions may cover any or all of personal belief and conduct, symbols and ritual, worldviews and ideologies, structures and institutions. In a review of theories within the sociology of religion, Beckford adopts an inclusive approach – religion is 'concern for the "felt whole" or for the ultimate significance of things' (Beckford 1989: 4). Beyer suggests similarly that 'religion [is that which] functions to lend meaning to the root indeterminability of all meaningful human communication and which offers ways of overcoming or at least managing this indeterminability and its consequences' (Beyer 1994: 6). These very broad approaches are the context within which most globalist theories of religion are formulated, with the result that the latter are particularly abstract.

Problems of definition indicate the remarkable flexibility of religious belief and practice. In a very limited sense religion can be a matter of personal experience; at any more general level, however, it is a social phenomenon that has a communal or corporate character. Religious traditions are invariably contested within and between social collectives – communities, classes, 'societies' – among which relations in general may be unstable. Religious references are constantly being appropriated and reappropriated: it is not a coincidence that the notions of sect and sectarianism have a religious association, bearing witness to processes of social construction and reconstruction which are especially significant at periods of rapid social change. As Moyser notes with reference to political relations:

> It is . . . a matter, in many contexts, of inter-group tensions and conflicts. Both solidarities and conflicts may have a religious element, revolving around shared or discrepant images of the sacred. But in addition they may also acquire other elements, either cultural (ethnicity, language, race etc.)

> or economic (wealth, occupation, class etc.) or a complex mixture of both. That is why, so often, inter-group 'religious' cleavages are complex divisions to analyse.
>
> (Moyser 1991: 11)

Specific 'linkages' of religion to other ideas, institutions and structures may vary enormously. In some circumstances religion may appear to operate as a mechanism of social control and integration; in others it may promote active rejection of relations of domination. Indeed, it may appear to sanction both domination and liberation within the same context, sometimes within the same institution.[10] What is required to comprehend the character of religion is to place it in the appropriate socio-economic, political and cultural contexts.

Global theory adopts a different approach. Robertson comments of cultural phenomena that 'uniqueness cannot simply be regarded as a thing-in-itself'. He goes on:

> It largely depends upon both the thematisation and diffusion of 'universal' ideas concerning the appropriateness of being unique *in a context*, which is an *empirical* matter, and the employment of criteria on the part of scholarly observers, which is an *analytical* issue.
>
> (Robertson 1992: 130, author's emphases)

What is unique or specific is to be understood as an expression of the relationship between the local, or particular, and the global, Robertson argues. In fact, most global accounts exclude the specific. Consistent with its emphasis on large issues, global theory strains to a higher level of generalisation and 'empirical matters' are rarely a concern. The result is that global perspectives on religion strongly discourage contextual understanding. Indeed, global accounts in general evacuate beliefs, practices and institutions of their specific significance, leaving 'religion' as little more than a reference to 'the whole', or to traditions of the supernatural, sacred, super-empirical, or transcendent – whatever definition is in use.

The functional model

Despite its references to the local or particular, global theory is preoccupied with abstract, often abstruse generalisation about structure and coherence. Such an approach is not original: it reasserts the functionalist conceptual framework that has dominated Western social theory since the late nineteenth century. Robertson's work, for example, is replete with references to functionalist sociology. Despite his claim to have modified Parsonian systems theory, with its emphasis on structure and integrity, Robertson's 'single place', 'unicity' and 'global field' echo the notions of organicism which have been at the heart of mainstream sociology.

Robertson insists that his approach allows for disjuncture and 'global disorder' (Robertson 1992: 188). He emphasises that stresses and tensions appear within his global field, largely as a result of the interaction between universal and particular elements.[11] None the less, on Robertson's account, all the elements of the global system operate within a globicity which is 'systemically' interconnected and throughout his work there is an unwritten suggestion that the global has an organic dimension within which tensions are accommodated.

Beyer's work is also marked by the Parsonian heritage. He acknowledges the work of Niklas Luhmann, a student of Parsons, who developed the latter's normative functionalism into a 'contingency functionalism' of inter-related but autonomous systems and sub-systems: economy, science, law, education, family and religion. Order and disorder are said to be the result of the interactions of these systems; their 'contingent adjustments' explaining much of the modernisation process and the emergence of the global (Luhmann 1982). In this context, Beckford explains, Luhmann thinks of religion as 'performing a necessary function on the level of the societal system' (Beckford 1989: 86). Hence Beyer's approach to religion as 'a type of communication which functions to lend meaning to . . . human communication' (Beyer 1994: 6).

The functionalist framework directs attention away from the specificities of socio-economic and cultural context. It demands that the particular be considered only in terms of an imagined and harmonious whole: hence the 'complementary' nature of globality and religion. One implication is that anti-systemic movements including religious activisms can be interpreted merely as a new engagement with the world system – affirmations of the link between the particular and the universal. It explains everything and nothing, emptying religious belief and religious activisms of all social significance. It has no explanatory value: the rise and fall of religious movements, their changing social and political agendas, and their internal inconsistencies and contradictions, remain mysterious. On this account movements such as Islamism and Liberation Theology might indeed be of supernatural origin.

Modernisation

The functionalist perspective is of a piece with modernisation theory and with modernisers' approaches to religion. As applied to the Third World during the high age of functionalist orthodoxy in the 1940s and 1950s, modernisation theory saw economic and political backwardness resolved by progress towards integration in a world economic system defined by the Western model. This, it was suggested, was not only necessary but was willed by the peoples of the Third World. As Daniel Lerner asserted in one regional context: 'What the West is . . . the Middle East seeks to become' (Lerner 1964: 47). The non-Western world could save itself, he argued, by adapting to the paradigm of modernisation.

To advance successfully, the modernisers argued, Third World societies must secularise, abandoning or at least modifying 'traditional' belief. Only when in progress from the Traditional pole towards the Modern could they energise their societies and find a place within the harmony of an expanding world (Lerner

1964; Rostow 1971). Secularisation, however, did not dictate abandonment of religion. Rather, it implied a move towards patterns of belief and practice associated with the West and in particular with the privatised North American Protestantism which constituted an ideal type of the modern society.

The aspirations and religious values of Middle America permeated modernisation theory. Lerner observed memorably in *The Passing of Traditional Society*: 'So long as a poor man laments his lot we are in the Traditional universe symbolised by Scriptural glorification of poverty and castigation of wealth' (Lerner 1964: 231). Rather than 'lament' poverty, the Third World should embrace the prosperity doctrines of the West, he maintained. Not to do so was perverse, a denial of progress and of the will of Third World peoples to become 'what the West is'.

Despite global theorists' absorption of much of postmodern, and indeed 'poststructural', theory, with its rejection of the 'grand narrative' of modernist progress, their view of religion remains within a functional and developmentalist perspective. Even sympathetic critics identify these difficulties. Kavolis comments that globalisation postulates 'a Durkheimian inevitability of moving, sooner or later, towards a universal value hierarchy' (Robertson 1992: 129) and Turner notes the criticism that global theory 'is evolutionary and teleological . . . in fact a new version of Westernisation' (Turner 1994: 108).

In this context it is worth noting Mottahedeh's comment on Huntington's theory of civilisational conflict. The idea of a global clash of religions, he observes, has given the United States of the 1990s 'what it most desires' – principles to make sense of the post-Cold War era. It has also offered 'a sense of purpose', a means of interpreting world events which gives meaning to existing economic and political relations at the world level (Mottahedeh 1995: 1). It is indeed 'a new form of Westernisation'.

Homogenisation

Global theory establishes all manner of encompassing categories. In the field of religion, the major world traditions are made both historically continuous and socially inclusive, cross-cutting economic, political and geographic boundaries. They are viewed as largely undifferentiated masses of believers interacting with the global system as abstract particularisms/universalisms or as units or sub-units within it. Thus we hear of the 'universal Church', 'global Catholicism' or 'the Islamic world'.

There is a special problem with religions centred in the Third World. The dichotomy of the West and 'the Rest' appears repeatedly, as in Huntington's theory of conflict between a unified 'West' and 'Muslim, Confucian, Hindu and Buddhist societies' (Huntington 1993: 45). The difficulty is particularly acute in the case of Islam and the notions of 'Muslim society' or 'global Islam'. Here the question of European self-image and the construction of the Other is an important historical legacy but also one reinforced by contemporary preoccupations: the polarisation of Modern and Traditional, Developed and Undeveloped, and most recently the concern with a perceived 'alien' presence outside/within Western society (Balibar 1991).

There is a very persistent view that religion is a primary element in the identity of individuals and groups within notionally 'Islamic' societies or communities. This idea is rooted in Orientalist notions of an essential Islam, in which Muslims are bearers of primal impulses absorbed from religious tradition. On this account Muslims are at *base* only 'religious' beings, sharing the atavism which is Islam and which promotes a desire to negate other cultures – hence the current Western media fantasy of a 'global *intifada*' (Esposito 1992: 182). Such a view continues to distract from the vast range of religious identification within and across societies in which Islamic traditions are influential: matters of place, class, gender, language, ethnicity, kinship, political affiliation and a host of other factors invariably subsume a specifically Muslim identity. In addition, as I wish to show, religion is not an historically continuous presence in individual or group identification but varies greatly in its character and influence over time.[12]

Once the global entity is assumed, however, a whole range of social and cultural forms may be placed within it. Even those who reject Orientalist prejudice are drawn into a homogenising practice: Turner, for example, writes systematically of 'the world of Islam' and of 'the global position of Islam' (Turner 1994: 12). These terms are meaningful only in so far as a vast range of people might perceive themselves *under specific and various circumstances* as 'Muslim'. There is no effective Islamic coherence – a matter which remains of overwhelming concern to many Islamic social and political movements. Notions of an inclusive world of believers, especially those absorbed by non-Muslims, invariably reproduce ideas about a harmonious community which are associated with particular, contested readings of Islamic tradition.[13]

A theme of contemporary Islamic activism is that of increasing *difference* within the collective, the *umma*. Believers are seen as divided by conflicts between ruler and ruled, between Muslim masses and 'external' and 'internal' colonialisms, between rival currents and sects, and between notionally Islamic nation-states.[14] This may indeed prompt 'a search for the fundamental' of unity in faith, producing renewed myths of a Golden Age and the aspiration to re-establish a homogeneous, unsullied community of believers – an important issue to which I wish to return. The quest for the *umma* does not, however, bear witness to a religious coherence which is 'the Islamic world', a notion which becomes another obstacle to understanding the dynamics of religious activism.[15]

Fundamentalism

A different aspect of the homogenisation problem lies in the identification of a global 'search for fundamentals' that easily becomes a construction of global 'fundamentalism'. For Robertson, for example, the varieties of religious 'fundamentalism' are singular and are nothing less than 'a product of globality' (Robertson 1992: 170). Here is an especially strong example of the global determinism embraced by some theorists and of the homogenisation by which it is accompanied.

As applied to religious movements, the notion of fundamentalism had its origins in North America in the early twentieth century, when it was applied to a specific interdenominational Protestant movement which insisted on the notion of inerrancy of the Bible (Hallencreutz and Westerlund 1996: 4–5). It has since been applied to all manner of religious activisms, most of which are in the non-Western world and have quite different agendas. Hallencreutz and Westerlund observe tersely that the habit of labelling Muslims and other non-Christians 'fundamentalist' may be obvious but that 'it is more difficult to find adequate scholarly reasons for it'. They comment: 'as a derogatory concept, tied to Western stereotypes and Christian presuppositions, it easily causes misunderstanding and prevents the understandings of the dynamics and characteristics of different religious groups with explicit political objectives' (*ibid.*: 4).

In the West the term has become a synonym for religious activisms, notably those of the Third World, and especially for Islamic currents.[16] Much interest in the West seems to be a response to what Lawrence calls the 'shock' of the Iranian revolution of 1979, the 'surprise' of the Khomeini phenomenon and the difficulty of coming to terms with Islamism in general (Lawrence 1990: ix). It is in this context that a 'Fundamentalist Studies' has emerged in the United States, institutionalised by the American Academy of Arts and Sciences in The Fundamentalism Project at the University of Chicago. Ostensibly a school of comparative studies with 'global' reach, the Project in fact focuses on the Middle East – its preoccupation with the region suggesting a neo-Orientalism in the mould of European institutions established in the nineteenth century for specialist analysis of 'the East' (Hallencreutz and Westerlund 1996: 2).

Some of the difficulties which arise when 'fundamentalism' is given global status can be observed in Lawrence's own book, *Defenders of God*. This analysis claims to discern a common project among American Protestants, Jewish settlers in Israel and Muslim activists in the Arab world – a 'revolt against the modern age' (Lawrence 1990). This extraordinary claim goes so far as to liquidate the contextual basis for understanding religious belief and activity: on this account any and all religious, social or political currents might be said to be globalisms. Shepherd comments: 'US fundamentalists are part of a society that was and is imperialist in various senses, and generally they share the attitudes that underlie imperialism', whereas Muslim 'fundamentalists' are 'part of societies that have suffered from imperialism and are anti-imperialist, often stridently so' (Westerlund 1996: 5). Although this comment also homogenises Muslim 'fundamentalism' (in practice some Muslim activists are far from hostile to the West),[17] Shepherd is right to maintain that the term obscures socio-cultural and political specifics which give each movement its character and its dynamic.

Human agency

For most of the twentieth century sociologists of religion have focused on what they saw as the linked issues of modernisation and secularisation. Religion, it was argued, was becoming increasingly individualised and privatised, and was less and less a framework or a point of reference for public action. The development which has now disturbed this orthodoxy is precisely the question of public engagement – the issue of religion and social action.

Religious activism turns on the construction of collectivities, of specific groups of agents who coordinate their actions in the light of an identity which they believe themselves to share. As Callinicos has pointed out, once a positive relationship between structural factors and social agents is allowed, there is 'enormous scope for empirical enquiry into the specific conditions favouring or impeding the formation of collectivities' (Callinicos 1995: 237). But just as globalist theory discourages examination of the specific social contexts in which religious movements develop, so it also narrows the scope for inquiry into the formation of the collective. First, it proposes a structure which constrains individual and collective relations within the whole, making sub-global structures largely irrelevant. Second, the human subject is marginalised. In the case of 'contingency functionalism', for example, subjects disappear entirely, operating only in the context of communication between systems and subsystems. As Luhmann puts it, 'the action system is the subject of the agents' (Beckford 1989: 80). Beyer observes that for Luhmann, 'society does not consist of human beings [sic] but of communications' (Beyer 1994: 58). Robertson, who insists that there is a place for the individual within his 'globicity', none the less comments: 'globalisation involves and continues to involve the *institutionalised construction* of the individual' (Robertson 1992: 105, emphasis in original).

The place of individual and collective subjects within this perspective worries even sympathetic analysts of the global account. In a recent review of global theories which is supportive of Robertson's work, Axford comments: 'the question of whether local subjects are not only influenced by global scripts but are able to influence them, has to be treated as an empirical matter and not one to be settled by mere stipulation' (Axford 1995: 3). But with empirical matters written out of globalist theory we are left with stipulation which in the case of religion is of little more value than believers' accounts of their own universalist perspectives.

Religious activism in the Third World

An analysis of religious activism which draws fully on empirical materials is not within the scope of this chapter. But even a brief examination of the major activisms brings to light key issues which the globalist account marginalises or ignores. I want to look at two movements which have had the greatest impact as activisms and which have stimulated most interest among global theorists – Liberation Theology and Islamism. In each case I want to examine the intimate association with social conflict, the importance of human agency, and the fluidity of religious ideas.

Liberation Theology

For twenty years from the late 1960s the fastest-growing social movement in Latin America was that associated with Liberation Theology. Its emergence shocked the church establishment and those who had seen Catholicism as a conservative, integrative force across the continent (Smith 1991: 3). Although its impact varied from country to country the net effect was so profound that it has been likened to the European Reformation of the sixteenth century (*ibid.* 4).

For centuries the church had been linked with local power structures, especially with owners of the *latifundia* – vast estates which dominated the rural economy. This relationship was highly structured and had its expression in a theology 'which perceived established hierarchies as an expression of the divine will' (Medhurst 1991: 191). Even the emergence of particularly stark patterns of uneven development did not disturb the structures of religious privilege. By the mid-twentieth century, urbanisation, industrialisation and huge movements of population were destabilising the old order. Increasingly, peasants, rural and urban workers and the dispossessed, entered struggles against those most closely linked to the ecclesiastical establishment. The latter responded by asserting the notion of the 'fortress Church' – ready 'to ward off threats from the surrounding society, pending its re-conquest' (*ibid.* 194).

The church began to lose ground to secular parties, especially to communist currents, and to more active Protestant groups. This pattern was particularly clear in rural areas and among the urban poor. Here the influence of the church had always been weaker: popular religious practice was often a syncretism that combined orthodox Catholic traditions with pre-Columbian elements and sometimes with indigenous African influences (Rowe and Schelling 1991). As a result, by the 1960s the religious establishment was coming under intense pressure from below: the response from some sections of junior clergy was to offer the church itself as a framework for social and political action. Increasingly, young priests became facilitators of movements for land reform, and of trade union and community action.

The Liberationists rapidly extended their influence. Their project was radical: borrowing ideas from secular currents including Marxism, they declared that 'the peoples of the Third World are the proletariat of today's humanity', that 'wealth must be shared by all' and that fundamental change was on the agenda (Smith 1991: 16). The sacred texts of the Judaeo-Christian tradition underwent a reassessment. Lehmann comments:

> These texts, this life [of Jesus Christ] and that tradition had been successively distorted and instrumentalised for oppressive purposes by feudalism, capitalism and now dependency and underdevelopment; the

> task was to recuperate them and turn them into the spiritual, ideological and practical basis for liberation. The new religious thought was radical in the strict sense that it proclaimed a return to the roots – to the roots, in this case, of Christianity.
>
> (Lehmann 1990: 118)

The implications for the church, Lehmann concluded, were 'little short of revolutionary' (*ibid.*: 120).

These developments, with their specifically Latin American character, were accelerated by events at the international level, especially by the success of anti-colonial movements such as those in Algeria and Vietnam. Groups of clergy began to interact with the secular left, producing political strategies which attempted to synthesise Marxism and Catholicism. In 1968 – the 'year of revolutions' – a meeting of senior clergy at Medellin, Columbia, formalised a new mission for the church across the region: it was to move from its 'fortress' *into* the masses, bearing a doctrine of liberation in which citizens of the Third World would address actively the problem of their oppression.

The crisis faced by the left across Latin America, which came to a head with the defeat imposed by the Chilean military coup of 1973, consolidated radical Catholic influence. Medhurst comments: 'In the face of repression, and in the absence of alternative channels for the articulation of dissent, the Church provided opposition groups with otherwise unavailable opportunities for rallying their forces' (Medhurst 1991: 207). During the 1970s the movement was able to mobilise on a massive scale, especially through ecclesiastical base communities (CEBs), groups of lay people under clerical leadership who came together for religious and educational purposes but which also mobilised for extensive social and political activity. By the end of the decade there were hundreds of thousands of CEBs involving millions of participants, the vast majority of whom were the urban poor, peasants and workers[18] (Smith 1991: 20).

The movement energised vast areas of the church: at the same time it created a profound institutional crisis. Lehmann comments that the space opened by the church was not just a political space, 'it was also [a space] in discourse, the possibility that a non-Marxist language of political agitation could develop' (Lehmann 1990: 96). In effect, sections of the church were reconstructing tradition as a means to express both protest and aspiration. The agency of change within the church was not the hierarchy itself, nor even its most liberal elements, but the energies of an increasingly impatient force from below. The 'space' created by the church was one which accommodated *their* specific demands and hopes. As Smith observes, the religious establishment had been compelled to adapt to an 'insurgent consciousness' (Smith 1991: 109).

Levine comments on the political implications of a movement which primarily engaged the dispossessed:

> The assertion by subordinate groups of a right to sacred power and a capacity for autonomous interpretation and action is a conflict-charged, and in this sense, a necessarily political act. Explicitly political motives

> may not be primary for this to be true [and] no matter, for such initiatives are viewed as political by the powers that be, who correctly understand the assertions of sacred power to be just a short and easy step away from efforts to acquire material power and to challenge authority in general.
>
> (Levine 1992: 321)

In fact much of the movement was self-consciously political and subversive. On its most radical wing, activists were weakening ties with local power structures and some priests even joined guerrilla struggles mounted by the secular left. Among the mass of activists differences between secular and religious programmes of action were often hardly discernible. The 'powers that be' responded with a storm of protest against 'Marxist infiltration' and 'red priests', and amid fear of loss of its historic links with the continent's rulers, the conservative core of the hierarchy organised a counter-movement. This succeeded in purging some radicals from positions of influence and eventually in modifying the church's formal position at a further conference in Puebla, Mexico, in 1979.

Growth continued into Central America, however, where by the late 1970s Liberationism was said to be 'massive and extremely widespread' (Booth and Walker 1993: 136). Alarmed by this development, and especially by the active involvement of clergy in the Sandinista revolution in Nicaragua, the United States began an urgent campaign of counter-activity. It found ready allies in a church establishment which had 'unleashed a tiger' it now officially sought to tame (Lehmann 1990: 130). At the same time the United States poured funds into conservative Protestant activisms which set out to organise against 'satanic' Liberation Theology (Diamond 1992: 53). By the mid-1980s, the Liberationists found themselves under relentless pressure. Constrained by their position within the religious establishment they were unable to offer unconditional support to continued mass activities. CEBs began to falter and 'radicalised' congregations to shrink, while there was a marked growth in Protestant and syncretic/Africanist currents.

Many radicalised clergy now operated as a brake upon hitherto dynamic areas of the movement. Their perspective had been one of working for the poor, what Leonardo and Clodovis Boff, two leading figures in the movement, called '"class conversion" . . . effective solidarity with the oppressed and their liberation' (Boff and Boff 1987: 23). This approach, despite the formal stress on mass self-activity, was consistent with the church's traditional authoritarianism and that of the very conservative social forces with which Catholicism had long been identified. As Pasara has pointed out, the radical Catholic style of those who went 'to the people' had strong elements of elitism and verticality (Levine 1992: 357). Levine notes that when the movement from below faltered, faced repression or competition from other activist currents, radical clerics were often left 'stranded, confused and exposed' (*ibid.*). They had not achieved the Boffs' 'class conversion' and an inherent authoritarianism often came into play, operating 'to weaken popular movements while energising and motivating right-wing opponents' (*ibid.*).

The radicalised clergy had ridden a rollercoaster of struggles initiated from below which they had attempted to give focus to within the Christian tradition. In the end they could neither free themselves from the masses nor, critically, from the church. By the 1990s, Smith comments, the movement's future depended upon how well it could adapt to changing conditions. Its fate was not to be determined by 'a band of [clerical] radicals' but by 'the uncontrollable social forces that make history' (Smith 1991: 237).

Global and globality

Liberation Theology has emerged within the turmoil induced by rapid economic, social and cultural change. The mass of participants have come from among those most subject to the insecurity and dislocation of modern life. The movement has not been limited to a region, country, ethnic or religious group but has had a pan-continental reach. It has been influenced by factors operating at a supranational level and has had its own wider impact: Liberation Theology conferences have been held in North America, Europe, Africa and Asia, and the Liberation philosophy has appeared in locations as diverse as the Philippines and South Africa (Frostin 1989: 65). Medhurst comments that as a result of the movement, 'the Latin American Church has made unprecedently significant contributions to perennial debates about the relationships of religion, society and politics' (Medhurst 1991: 214).

In these contexts Liberation Theology needs to be understood at the level of world affairs. To see it as an expression of 'globality', however, is to rob the movement of the qualities which make it such a vigorous activism. Liberation Theology is not merely an 'antisystemic element' or a sub-unit of an imagined 'unicity', as in the globalist account. Nor does it fit within the functionalist models which dominate global theory. Rather, it is a dynamic response to systemic contradiction, a testimony to disjuncture within Latin American society. As Smith makes clear, the movement emerged 'concurrent with [mass] forces' which sought a radical solution to the crises precipitated by uneven development across the continent (Smith 1991: 236). In this sense, the movement has been above all a positive assertion of human agency.

The Latin American movement is an example *par excellence* of religious traditions undergoing modification by social actors who have brought new meanings to scriptural orthodoxy. Lehmann's 'return to the roots' is indeed a search for fundamentals but one undertaken by those with specific aspirations for change which are inseparable from the structural factors which constrain them at the levels of their own experience. A whole range of collectives formed among the exploited and the dispossessed has provided an environment within which radical Catholicism could emerge. As an intellectual current, it has been marked by the need to generate 'insurgent consciousness' adequate to engage with an impatient mass movement.

The Latin American case illustrates the flexibility of religion and its social constructedness. It shows how, on the one hand, religion can be closely connected to structures of privilege and can operate ideologically in the classic Marxian sense by making inequality 'an expression of the divine will'. On the other hand, religious traditions and even institutions can provide an idiom for expression of mass aspirations. In these processes human agents are decisive, both as aspirants to change and as those whose structural position makes them hostile to change, for Levine's 'powers that be' also represent human agents with a collective concern about sacred power. Only an empirical inquiry into the balance between these pro- and anti-systemic forces (and those who vacillate between them) allows us to understand the trajectory of the movement.

On the global account, religion is increasingly significant, even 'determinative', at the world level. With its universal references, religion is made to correspond to a new totality of human experience. But examination of the Latin American movement shows that the importance of religious ideas varies greatly – that they may be more significant or less significant for social agents under differing circumstances. This pattern also becomes clear when we examine Islamic activism.

Islamic activism in Egypt

The growth of Islamist movements is invariably seen as confirmation of the globalist thesis. In a recent introduction to global theories, Waters, for example, makes 'Islamic fundamentalism' one of 'the impersonal forces' driving globalisation (Waters 1995: 2). I have pointed out the dangers in such homogenisation. Here I want to look at two specific experiences: in Egypt – where the movement has had its longest influence; and in Iran – where it has had the most profound impacts.

Islamism long predates its recent rise to prominence. Its roots are in the response of some sections of the religious establishment, the *'ulema*, to the rapid penetration of the Middle East by a rampant European colonialism. In the mid-nineteenth century junior clerics in the Sunni Muslim regions began to reject the policies of imitation of Europe which had been favoured by local rulers and loyalist *'ulema*. Calling for a defence of Islamic culture, they asserted foundational religious values as the most appropriate means of countering the European threat.[19]

Their initial influence faded and by the early twentieth century secular nationalism emerged as the most vigorous political movement across the Middle East. By the 1920s, however, with the nationalists unable to stem Western advance, Islamism re-emerged in Egypt, this time as a mass movement in the form of the Muslim Brotherhood. Zubaida sums up its project as 'displacement of the existing order with one based on Islamic law and principles of social justice [achieved] through popular organisations and mobilisation' (Zubaida 1989: 155). The Brotherhood's vision was one of an Islamic society recuperated through mass political action. It attacked 'external' colonialism (European rule) and 'internal' colonialism (the influence of nationalist politicians and 'collaborationist' *'ulema*) (Mitchell 1969: ch. 8). The main aim was to expel the alien (British) presence and to implement *shari'a* as the basis of the legal system.

By organising a network of independent associations, community groups and trade unions, the Brotherhood became the first mass organisation in the Arab world. By the 1940s it had an estimated 500,000 members in Egypt and constituted the main threat to British rule. The movement faltered, however, especially as trade union and communist organisations emerged which proposed more radical programmes for change and soon captured their own mass support. The Islamist leadership retreated towards accommodation with the regime and its support began to ebb. In 1952 the army took power and within a few years the Brotherhood was in steep decline. For the next twenty years it operated at the margins of Egyptian society, while the wider influence it had anticipated throughout the Middle East failed to materialise.

In the late 1960s the military regime entered a deep crisis and the Islamist movement again grew with extraordinary speed. Throughout the 1970s Egypt was in ferment: the vision of an independent Third World economy had collapsed, a pioneering programme of economic 'liberalisation' was widening social inequalities,[20] and millions were migrating for work abroad. With secular radical currents largely ineffective (communists having earlier embraced the nationalist regime), a new Islamism emerged – a more insistent radicalism that demanded 'revolution' and immediate implementation of an Islamic order. Its mass appeal lay in the promise to lessen inequality, to attack state corruption and to bring stability and social justice.

The new movement's vision turned around revitalisation of the *umma*. Its ideologues pressed for immediate removal of secular and syncretist influences which, they maintained, prevented realisation of the Muslim collective. They elaborated a perspective which obliged all believers to struggle for removal of their rulers, a first step towards reassertion of the *umma* by means of a programme of radical *shari'a* reforms.

Drawing on the work of the most politically engaged figures of the Islamic tradition, leading figures of the movement such as Sayid Qutb shaped an interpretation of the past which could be a reference point for contemporary activism. They spelt out a novel politics – what Al-Azmeh describes as 'the precise and imminent interpretation of the pristine model' (Al-Azmeh 1993: 99) which made the original *umma* of the Prophet realisable within contemporary society. Ayubi sums up how Muhammed 'Abd al-Salam Faraj, a leading figure of the movement, argued for *jihad* (exertion) against the 'unbelieving' state:

> As is often the case [among radical Islamists] Faraj achieved his extremely militant formulation through a process of selective interpretation of the original texts. Only the militant verses of the Quran are quoted, only the 'combative' definitions of *jihad* are adopted and only the most stern jurists are referred to.
>
> (Ayubi 1991: 144)

Drawing in large numbers of urban poor, marginalised peasants and members of a bitterly frustrated middle class, organisations such as Al Jihad and *al-gama'at al-islamiyya* (the Islamic groups) proved highly destabilising to the Egyptian state. But their authoritarian style, based on the religious injunction rather than mass mobilisation, inhibited the movement. At the same time, many activists found that their engagement with the state constrained their political efforts to combat it.[21] Developing a strand of the Brotherhood's earlier strategy they attempted to overwhelm the state by building conspiratorial networks within the armed forces – an approach Ayubi sees as construction of 'a mirror image' of the existing polity (Ayubi 1991: 154).

Following the assassination of President Sadat in 1981, the government launched a campaign of repression which continued into the 1990s, varying in intensity as the Islamists' support waxed and waned. At various times the state has declared the movement to have been eradicated, only to face a resurgence of activity which, as in the early 1990s, has required a massive offensive to destroy a new popular base.

The Iranian experience

The Iranian revolution of 1979 is seen as a decisive moment in the rise of Islamism and in the intensification of globalising processes. According to Beyer, for example, the 'Islamic revolution'[22] needs to be seen as 'a direct response to globalization' (Beyer 1994: 160). It has often been seen as the outcome of cultural processes at work across the 'Islamic world'; in fact, the Iranian movement has a distinct and very specific history.

Islamism had a first impact in the 1890s but then became entirely marginal within Iranian society. Throughout the first half of the twentieth century, while Egypt was witnessing massive struggles under Islamist influence, almost the entire Iranian religious establishment advocated a quietism which amounted to endorsement of the secular regime.[23] It was not until the 1960s, when Iranian society began to go through processes of accelerated change, that a small group of *'ulema* around Khomeini began publicly to express opposition to the regime. At the same time, some lay figures close to the religious establishment but who were also under left-wing influence attempted a synthesis of Islamic and secular radical ideas that has sometimes been called an 'Islamic Marxism' (Abrahamian 1989).

During the 1960s and 1970s Iranian society went through a severe dislocation. Huge volumes of money entered the economy from the oil sector, intensifying commercialisation in the countryside, dispossession of the peasantry and mass migration to the cities. The state promised a restless population widespread social change: when by the late 1970s it had failed to deliver, a mass movement of almost unprecedented depth and energy emerged. For decades, national politics had been influenced by strong liberal and communist currents but these were now weak and disorganised and a network of junior clerics around Khomeini emerged as the most intransigent opposition to the regime.

The clerics advanced an 'Islamic strategy' that owed as much to the immediate demands of the masses as it did to Shi'a tradition. Using language that would have been unthinkable in an earlier period, they insisted that both *'ulema* and Muslim masses must fulfil a divine duty by overthrowing the Pahlavi regime. Khomeini argued that the notion of a separation between religion and politics, and the idea that Islamic scholars should not intervene in social and political affairs, had been 'formulated and propagated by the imperialist; it is only the irreligious who repeats them' (Moaddel 1993: 128). He demanded mass action against the Pahlavi state.

Khomeini's injunction for activism was publicly linked with struggle for a specific ideal: an Islamic state presented in highly populist terms that placed Khomeini alongside the radicals who had attempted to fuse socialist and Islamic traditions. He proposed an 'Islamic republic' which echoed the ideas of the mass movement:

> The weak will assume the leadership overall. God's promise will soon be fulfilled and the downtrodden will supplant the rich. In an Islamic republic there is no oppression and injustice, there are no rich and poor, everyone will have equal rights. In an Islamic republic all the layers of society, all religions, all races and communities will have equal rights.
>
> (Engineer 1994: 181)

The effect, comments Zubaida, was to suggest 'commitment to democracy and social justice [which] led many democratic and left forces to acquiesce in Khomeini's leadership' (Zubaida 1989: 60). With these secular alternatives already disoriented, what Khomeini called 'the progressive dictates of Islam' became a focal point within the mass movement (Khomeini 1981: 244).

After the fall of the regime in 1979 the mass movement expanded into what Bakhash has called 'a riot of participatory democracy' (Bakhash 1985: 56). Several ethnic groups declared autonomy, vast areas of land were seized by the peasantry, most of manufacturing and service industry came under the control of elected workplace councils, and local committees took on the running of community affairs (Bakhash 1985: ch. 3; Bayat 1987). Khomeini and his closest supporters launched a ferocious assault on these movements – on those who wished to continue the process of change or who questioned the new polity, which was emerging as a highly authoritarian, centralised clerical rule. The Islamic agenda, it became clear, would not be set by the mass movement or the theorists of 'radical' Islam but by the ayatollah and his immediate supporters, whose reading of the *shari'a* was that favoured by particularly conservative sections of Iranian society, notably mercantile layers with which the *'ulema* had long had close links.[24]

With the decade of the 1980s dominated by war with Iraq, the Iranian regime stabilised as a nationalism strongly coloured by Islamic rhetoric which ruled over a resentful but largely passive population. As Levine has commented of the Latin American case, 'the concept and conduct of authority remain[ed] wedded to hierarchy' (Levine 1992: 351).

The global dimension

It is easy to see why Islamism has been associated with globalism. Muslim activists' vision of a pan-Islamic *umma* which might link believers across national boundaries, ethnic and political barriers has the ring of a universalism in keeping with both the 'compression' of the modern world and the apprehension of a totalising socio-cultural environment. Its strong assertion of scripturalism and legalism also appears to offer certainties which answer the quest for fundamentals among those who face the greatest uncertainty in a world of rapid change.

Islamism has also had a wide impact, especially since the Iranian revolution. The 1979 upheaval was one of the first 'global' media events. International television and press crews had access to many of the scenes of political activity and these were filmed and broadcast worldwide. Sreberny-Mohammedi and Mohammedi observe that, as in the case of other recent political upheavals, instant dissemination around the world of such images 'fosters a possible "contagion effect"' (Sreberny-Mohammedi and Mohammedi 1994: 28). Walid Abdelnasser describes its impact on Muslim activists elsewhere:

> For them, the Iranian revolution represented a new reality to which all Muslims could turn and relate their history, ideals and the development of the entire Islamic *umma*. The revolution meant that in practice, Islam could win against imperialist powers and oppression.
>
> (Abdelnasser 1994: 67)

Following the Iranian events, Islamist movements grew rapidly in the Arab world and appeared in some regions of Africa and Central Asia. In a series of countries activists gained confidence to attempt similar assaults on state power, producing anti-Islamist panics in Arab capitals and in both Washington and Moscow. These were not novel developments, however. They echoed events in the 1940s and 1950s, when mass enthusiasm for secular nationalism spread across the Middle East, culminating in the Iraqi revolution of 1958, when the United States declared the region 'the most dangerous place in the world' (Hazelton 1986: 26).

Whereas radical nationalism took power in states across the Middle East, Islamism succeeded after the Iranian revolution only in Sudan. What is striking about the fate of movements which have sought to repeat the Iranian experience is their extreme volatility and ultimate lack of success. The Algerian case provides an ideal type of the pattern of the 1980s and 1990s: a mass engagement followed by swift retreat. This rise and steep fall is taken by a number of analysts to be an index of deep contradiction within the movement as a whole: what Roy calls 'the failure of political Islam' (Roy 1994).

The volatility of the Islamist movement again brings the globalist account of contemporary religiosity into question. Islamism has not been a progressively assertive force, drawing believers into a pan-religious vision which expresses 'globicity'. Nor does it show the perceived determinism associated with globality, as Waters maintains in viewing the movement as one of the global forces 'beyond the control and intentions of any individual or group of individuals' (Waters 1995: 2).

Islamism is a series of currents, varying in influence over the course of more than 100 years, which has advanced and retreated under a great variety of circumstances. For long periods it has been marginal within nation-states and across Muslim communities and sects. Increased political significance and greater social and cultural influence has come only when a rise in general levels of mass activity has coincided with a retreat of secular currents, especially of radical nationalism and of the left. This is clearly discernible in the Egyptian and Iranian cases and holds good for almost every case of Islamic 'resurgence' across the Middle East region.

Islamism shows the extraordinary elasticity of religious ideas. In Egypt, the movement has passed through several phases, emphasising different aspects of religious tradition. The movement of the 1930s was primarily a theology of anti-colonial struggle; that of the 1970s was a focused anti-statism which has been seen as an Islamicised Leninism (Roy 1994: 3). In Iran the *'ulema* were passive for generations. What brought them into political activity was a growing hostility to the state on the part of the mass of Iranians, combined with an awareness that the deficiencies of liberalism and of the communist currents had rendered secular politics ineffective. Khomeini's 'Islamic agenda' for the revolution was a world away from his years of acceptance of the status quo, although it was constructed from the same fund of traditions which had made senior *'ulema* ideologists for generations of Iranian rulers. As a number of writers have shown, Khomeini's dissimulation allowed him to seize leadership of the mass movement opportunistically and with great political skill (Engineer 1994: ch. 6; Irfani 1983: ch. 9). This was accomplished by making the aspirations of the masses his own – by giving them divine endorsement, albeit temporarily. As the mass movement retreated, the clerics' own agenda was revealed as a conservatism which left the structures of the Iranian state largely untouched (Roy 1994: ch. 8). (It is worth noting that for all the globalist assertions of a borderless world in which nation-states are rendered ineffective, it is the local state which has been the focus of Islamist concerns. In addition, the state has been an instrument of repression against Islamist insurgencies and by Islamist leaderships against religious activisms that threaten to move beyond their control.)[25]

The intensity of conflict in post-revolutionary Iran had its own negative impact on Muslims elsewhere. By the early 1990s, with Khomeini dead and the Iranian regime preoccupied with its own survival, the Iranian model had become far less influential. Its declining significance may be seen as one factor contributing to a general contraction in influence of Islamism as a wider movement.

Conclusion

Liberation Theology and Islamism have their specificities – their own complex interactions of classes, communities, traditions and distinct ideological currents. The most important difference between them as broad movements is in the attitude to mass self-activity, with the Latin American case showing a much fuller popular engagement than the halting, more authoritarian examples from the Middle East. But they do have a crucial similarity, for each has emerged within the context of struggles of the dispossessed. The two movements are marked by the experience of their participants as marginals within the dominant economic and political order. In this context, Haynes comments, 'important concerns of Iranian revolutionary dogma [sic] were also those of liberation theology' (Haynes 1993: 30). It is this challenge to hegemonic economic, political and social relations which has made Islamism a 'shock' and radical Catholicism a 'storm' sweeping Latin American society (Smith 1991: 21). The destabilisation associated with these movements has prompted strong reactions among the dominant world powers, prompting their ideologues to develop theories which both essentialise cultures and 'civilisations', and suppress notions of difference between rich and poor, ruler and ruled, or First and Third Worlds.

Not all religious movements are activisms and not all movements in the Third World are anti-systemic. Protestant evangelicalism, for example, has greatly increased its influence, partly as a result of vigorous efforts made by some Western states and their allies within the Third World to 'seed' and to sustain a current which stresses inner-worldliness, spirituality and political conservatism (Diamond 1992; Martin 1993). Evangelicals or Pentecostals do not provoke the anxiety which has accompanied growth of Christian radicalism or Islamism, nor warnings of a threat to global order. It is the *activisms* which have prompted a surge of interest in religion and politics worldwide – what Moyser describes as a sharp move from neglect to 'intense scrutiny' (Moyser 1991: ix).

Levine has commented on religion's transformative powers, arguing that they stem 'above all from its ability to cloak any activity with symbolic significance in ways that mobilise individual and group resources' (Levine 1992: 351). Most informed analyses of the major religious movements in the Third World reach a similar conclusion. Sometimes religious traditions are mobilised in ways which complement collective activity: here religion can facilitate change. In other cases, religious traditions remain closely tied to clerical establishments or other institutions with their own links to structures of privilege, and innovation may be inhibited. Only an investigation of the dynamic of each movement reveals the extent to which religion remains a point of reference for activism.

These investigations are inhibited by globalist theory, which gives religion a semi-autonomous status from which it generates influence across an imagined unicity. The patterns of rapid, even traumatic, change associated with Liberation Theology or Islamism largely disappear in the global account. The great fluidity of religious ideas is masked by an interpretation which makes their universalising potential a dimension of the 'global experience'. Here, as Turner has noted, globalism suggests an evolution towards final status in which religion is 'realised' in a unified world, projecting a particular developmentalist logic onto a vast range of religious ideas and movements.

Models which assume unilinear change, make social conflict pathological and strongly emphasise the totalising qualities of religion obscure the character of modern activisms. The level of abstraction at which they operate not only directs attention away from their specific dynamics but makes them weak predictive models. As with conventional approaches to secularisation criticised by Levine, 'They are [like] the functionalist sociological models whose stress on harmony and cohesion utterly failed to anticipate or explain the rise of the civil rights movement in the United States in the 1950s, not to mention the protests of contestation of subsequent decades' (Levine 1992: 352).

Notions of social cohesion sit uneasily in a world marked by increasing unevenness, tension and disorder. This is especially so in the true crisis areas of Latin America, Africa and Asia. Here there are likely to be further intense struggles for survival which will affect local power structures and wider relations of domination. Some struggles will take the form of religious activisms with radical agendas. They will rise, fall and reappear in new forms, with divine missions made on the basis of remodelled traditions. Globalist theory will not assist our understanding of such movements. Its unified world model will provide no explanation for their vigour or for the anxiety they create among local rulers and the dominant world powers. It will in fact deny the efforts of the dispossessed to change the relations of inequality which shape their lives – to seek a role in the making of their future.

Notes

1. Writing in 1958, Lerner endorsed the observation of some Islamist scholars that, confronted by the West's 'rationalist and positivist spirit', Islam was in rapid retreat (Lerner 1964: 45).
2. Here I use Islamism as a synonym for Islamic activism, a movement oriented on social and political change which mobilises on the basis of Islamic traditions.
3. O'Hagan comments that it has stimulated debate 'both in academic circles and in the corridors of power' (1995: 19). See O'Hagan (1995), Tarock (1995), Mottahedeh (1995) and Holmes (1997).
4. Various writers, notably Toynbee, have interpreted historical change as an expression of 'civilisational' developments (Federici 1995; Gowilt 1995). Axford observes that in the late twentieth century the term carries 'repugnant overtones about the superiority of one culture over others' (Axford 1995: 191). Some global theorists have none the less given it a new, world-encompassing significance.
5. Robertson writes of the importance of 'religioculture' in the global context (Robertson 1992: 96). Waters concludes that the current revitalisation of religion is inseparable from global processes (Waters 1995: ch. 6) and Beyer maintains that religion has become 'instrumental' in the development of globalisation (Beyer 1994: 3).

6. Beyer argues that it may become increasingly difficult to retreat into the isolation in which particular values can be confidently reasserted. He greatly understates, however, the potential for development of inward-looking movements which, like Utopianisms of the nineteenth century, seek to establish self-sufficient communities which are relatively isolated from the wider society. The Islamist sects of the 1960s and 1970s provide a particularly good example (Kepel 1985: ch. 5).
7. Turner, who shares much of this perspective, sums up how globalising developments have affected Islam and Christianity. Each had conceptualised themselves as world religions, although in the pre-modern period, he argues, each had a problematic relationship with local political systems, being unable to 'realise themselves' globally. The possibility of achieving such a final status, however, has been facilitated by the changes associated with modernity (Turner 1994: 83 and ch. 6).
8. Writers such as Abu-Lughod maintain that a 'world system' emerged as early as the thirteenth century in the form of commercial networks focused upon cities of the Middle East/Central Asia and within which Europe had only peripheral influence (Abu-Lughod 1989). But all-embracing processes of integration emerged relatively recently and are clearly associated with developments inextricably linked with 'modernity', notably with industrial capitalism and the rise of the nation state. Robertson is one of the few writers to attempt to 'periodise' this process, arguing convincingly that its key moment was in the late nineteenth century, when national and transnational linkages developed rapidly (Robertson 1990).
9. Hirst and Thompson observe that even the most optimistic scenarios for economic growth outside the core economies of the world system suggest that 'the Third World will remain marginal' (Hirst and Thompson 1996: 111).
10. As when sections of the Roman Catholic Church have both opposed and embraced Liberationism, or when the Islamic *'ulema* have argued for both political quietism and radical change.
11. Turner tellingly comments that in Robertson's analysis it is 'the conventional Hobbesian problem of order' that becomes 'necessarily a problem of global disorder' (Turner 1994: 111).
12. Islamic currents such as the movement for the Islamisation of knowledge would reject such comments as typical of Western analyses which seek to fragment the collective of believers. Such a view is itself associated, however, with the perspectives of particular sections of notionally Islamic societies.
13. Although some Islamic currents would argue that the community of believers – the *umma* – is actualised by the mere presence of Muslims, others maintain that a particular current, sect or movement represents the community and that others are in effect *kafir*, or non-believers. For a discussion of how contemporary Islamists have viewed these issues see Ayubi (1991: ch. 6).
14. Such critiques of the divided *umma* in a world of nation-states first emerged in the late nineteenth century in the work of Jamal al-Din al-Afghani. For an account of his work see Keddie (1968, 1972), Al-Husry (1980) and Moazzam (1984). On the later positions of the Muslim Brotherhood, see the work of Hassan al-Banna (Mitchell 1969). On recent 'radicals', see the arguments of Mawdudi in Mawdudi (1955) and Nasr (1994); on Qutb see Qutb (1988), Ayubi (1991) and Tripp (1994).

15. The practice of homogenising societies outside Europe and North America has not been confined to modernisation theory and related currents. The 'underdevelopment' theorists of the 1960s and 1970s, who strongly influenced radical nationalist movements in Latin America, Asia and the Middle East, similarly homogenised nation-states and whole continents, depicting the exploitation and oppression of an undifferentiated Third World. For a critique see Kiely (1995: ch. 3).
16. In a special report on Islam and the West, for example, *The Economist* described European approaches to Islam as expressed by a 'fundamental fear' of societies of the Middle East (6 August 1994).
17. Some Islamist currents have combined rhetorical opposition to 'imperialism' with close collaboration with Western powers at the political level – for example, in Afghanistan. Other currents have attempted to integrate into Western-dominated economic structures. See Zubaida (1989: ch. 6).
18. There were an estimated 150,000 to 200,000 CEBs (Smith 1991: 20). In the mid-1980s one archdiocese alone – that of Vitoria in Brazil – counted 50,000 members of CEBs, while in the city of Recife there were 269 groups and 70,000 participants (Lehmann 1990: 136).
19. The most effective early Islamist current was that of pan-Islam. Its key ideologue, Jamal al-Din al-Afghani, was active in Egypt, Iran and India during the last two decades of the nineteenth century. For a full account of Afghani's ideas and activities, see Keddie (1972).
20. President Sadat's programme of *infitah*, or 'opening', which anticipated by a decade the neo-liberalism espoused by governments and by transnational agencies such as the International Monetary Fund.
21. Many Islamists were army officers, professionals, teachers and administrators whose employment, housing and minor privileges depended upon the apparatus of state itself. Some of the contradictions which emerged are spelt out in Ayubi (1991). See also Davis (1984) and Harman (1994).
22. The idea of an 'Islamic revolution' is contentious. The movement of 1978–9 was a mass popular uprising in which Islamic currents were only one element. See Abrahamian (1982), Bakhash (1985) and especially Bayat (1987).
23. Contrary to the idea which has absorbed many analysts, Shi′a traditions have not induced leading clerics to challenge for state power. As Enayat has spelt out in the most comprehensive review of Islamic political thought, although Shi′a traditions are indeed a *potential* tool of radical activism, 'throughout the greater part of Shi′i history [this] never went beyond the potential state, remaining in practice merely a sanctifying tenet for the submissive acceptance of the *status quo*' (Enayat 1982: 25).

    The only clear example of sustained clerical activity over decades of quietism during the twentieth century was that of the short-lived Fedayan-e Islam (Richard 1983).
24. Zubaida observes that 'a fundamentalist revolution was achieved with the full support of secular democratic forces who were later to become its victims' (Zubaida 1989: 60). He might have added that Khomeini also initially received the backing of the mass of Islamic activists, among whom many were also later ruthlessly suppressed.

25. This points up an unresolved problem in Robertson's work. His 'national societies' are invariably presented as unities or coherences in which conflict is largely irrelevant. If his 'societies' are defined as nation-states it becomes impossible to evade the issue of power and the analysis of the state as an instrument of domination (see Robertson 1992).

References

Abdelnasser, W. M. (1994) *The Islamic Movement in Egypt: Perceptions of International Relations 1967–1981*, London: Kegan Paul International.

Abrahamian, E. (1982) *Iran Between Two Revolutions*, Princeton: Princeton University Press.

—— (1989) *Radical Islam: The Iranian Mojahedin*, London: I. B. Tauris.

Al-Azmeh, A. (1993) *Islams and Modernities*, London: Verso.

Al-Husry, K. A. (1980) *Origins of Modern Arab Political Thought*, New York: Caravan.

Appadurai, A. (1990) 'Disjuncture and difference in the global cultural economy', in M. Featherstone (ed.) *Global Culture*, London: Sage.

Axford, B. (1995) *The Global System: Economics, Politics and Culture*, Cambridge: Polity.

Ayubi, N. (1991) *Political Islam: Religion and Politics in the Arab World*, London: Routledge.

Bakhash, S. (1985) *The Reign of the Ayatollahs*, London: Unwin.

Balibar, E. (1991) 'Es Gibt keinen Staat in Europa: racism and politics in Europe today', *New Left Review* 186.

Bayat, A. (1987) *Workers and Revolution in Iran*, London: Zed.

Beckford, J. A. (1989) *Religion in Advanced Industrial Society*, London: Hyman.

Beyer, P. (1994) *Religion and Globalisation*, London: Sage.

Boff, L. and Boff, C. (1987) *Introducing Liberation Theology*, Tunbridge Wells: Burns & Oates.

Booth, J. A. and Walker, T. W. (1993) *Understanding Central America*, Boulder, CO: Westview.

Callinicos, A. (1995) *Theories and Narratives: Reflections on the Philosophy of History*, Cambridge: Polity.

Davis, E. (1984) 'Ideology, social class and Islamic radicalism in modern Egypt', in S. Arjomand (ed.) *From Nationalism to Revolutionary Islam*, Basingstoke: Macmillan.

Diamond, S. (1992) 'Holy warriors', in J. P. N. Pieterse (ed.) *Christianity and Hegemony: Religion and Politics on the Frontiers of Social Change*, Providence: Berg.

Enayat, H. (1982) *Modern Islamic Political Thought*, Basingstoke: Macmillan.

Engineer, A. A. (1994) *The Islamic State*, New Delhi: Vikas.

Esposito, J. L. (1992) *The Islamic Threat: Myth or Reality?* New York: Oxford University Press.

Featherstone, M. (1990) 'Global culture: an introduction', in M. Featherstone (ed.) *Global Culture*, London: Sage.

Federici, S. (1995) 'The God that never failed: the origins and crises of Western civilization', in S. Federici (ed.) *Enduring Western Civilization*, Westport: Praeger.

Frostin, P. (1989) 'The theological debate on liberation', in P. Katjavivi, P. Frostin and K. Mbuende (eds) *Church and Liberation in Namibia*, London: Pluto.

Giddens, A. (1990) *The Consequences of Modernity*, Cambridge: Polity.

GoGwilt, C. (1995) 'True West: the changing idea of the West from the 1880s to the 1920s', in S. Federici (ed.) *Enduring Western Civilization*, Westport: Praeger.

Gupta, B. S. (1986) *Afghanistan: Politics, Economics and Society*, London: Pinter.

Hall, S. (1992) 'The question of cultural identity', in S. Hall, D. Held and A. McGrew (eds) *Modernity and Its Futures*, Cambridge: Open University/Polity.

Hallencreutz, C. F. and Westerlund, D. (1996) 'Anti-secularist policies of religion', in C. F. Hallencreutz and D. Westerlund (eds) *Questioning the Secular State: the Worldwide Resurgence of Religion in Politics*, London: Hurst.

Harman, C. (1994) 'The Prophet and the Proletariat', *International Socialism* 64.

Harman, C. (1996) 'Globalisation: a critique of the new orthodoxy', *International Socialism* 73.

Harvey, D. (1989) *The Condition of Postmodernity*, Oxford: Blackwell.

Haynes, J. (1993) *Religion in Third World Politics*, Buckingham: Open University Press.

Hazelton, F. (1986) 'Iraq to 1963', in CARDRI, *Saddam's Iraq*, London: Zed.

Hirst, P. Q. and Thompson, G. (1996) *Globalization in Question*, Cambridge: Polity.

Holmes, S. (1997) 'In search of new enemies', *London Review of Books*, 24 April.

Hoogvelt, A. (1997) *Globalisation and the Postcolonial World*, Basingstoke: Macmillan.

Huntington, S. (1993) 'The clash of civilizations?', *Foreign Affairs* 72 (3), Summer.

—— (1997) 'The West and the world', *Foreign Affairs* November/December 1996.

Hussain, A. (1984) 'The ideology of orientalism', in A. Hussain, R. Olson and J. Qureshi (eds) *Orientalism, Islam and Islamists*, Brattleboro: Amana.

Irfani, S. (1983) *Revolutionary Islam in Iran*, London: Zed.

Keddie, N. R. (1968) *An Islamic Response to Imperialism: Political and Religious Writings of Sayyid Jamal ad-Din 'al-Afghani'*, Berkeley: University of California Press.

Keddie, N. R. (1972) *Sayyid Jamal ad-Din 'al-Afghani': A Political Biography*, Berkeley: University of California Press.

Kepel, G. (1984) *The Prophet and Pharaoh: Muslim Extremism in Egypt*, London: Al Saqi.

Khomeini, I. (1981) *Islam and Revolution*, Berkeley: Mizan.

Kiely, R. (1995) *Sociology and Development*, London: UCL Press.

Lawrence, B. B. (1990) *Defenders of God: the Fundamentalist Revolt Against the Modern Age*, London: I. B. Tauris.

Lehmann, D. (1990) *Democracy and Development in Latin America*, Cambridge: Polity.

Lerner, D. (1964) *The Passing of Traditional Society: Modernizing the Middle East*, New York: Free Press.

Levine, D. H. (1992) *Popular Voices in Latin American Catholicism*, Princeton: Princeton University Press.

Luhmann, N. (1982) *The Differentiation of Society*, New York: Columbia University Press.

Martin, D. (1993) 'The evangelical expansion south of the American border', in E. Barker, J. A. Beckford and K. Dobbelaere (eds) *Secularization, Rationalism and Sectarianism*, Oxford: Clarendon.

Medhurst, K. (1991) 'Politics and religion in Latin America', in G. Moyser (ed.) *Politics and Religion in the Modern World*, London: Routledge.

Mitchell, R. P. (1969) *The Society of the Muslim Brothers*, London: Oxford University Press.
Moaddel, M. (1993) *Class, Politics and Ideology in the Iranian Revolution*, New York: Columbia University Press.
Moazzam, A. (1984) *Jamal ad-Din al-Afghani: a Muslim Intellectual*, New Delhi: Concept Publishing Company.
Morley, D. and Robins, K. (1995) *Spaces of Identity*, London: Routledge.
Mottahedeh, R. (1995) 'The clash of civilizations: an Islamicist's critique', *Harvard Middle Eastern and Islamic Review* 2(2).
Moyser, G. (1991) 'Politics and religion in the modern world: an overview', in G. Moyser (ed.) *Politics and Religion in the Modern World*, London: Routledge.
O'Hagan J. (1995) 'Civilisational conflict? looking for new enemies', *Third World Quarterly* 16(1).
Ohmae, K. (1995) *The End of the Nation-State: the Rise of Regional Economies*, New York: HarperCollins.
Pieterse, J. P. N. (ed.) (1992) *Christianity and Hegemony: Religion and Politics on the Frontiers of Social Change*, Providence: Berg.
Qutb, S. (1988) *Milestones*, Karachi: International Islamic Publishers.
Richard, Y. (1983) 'Ayatollah Kashani: precursor of the Islamic Republic?', in N. R. Keddie (ed.) *Religion and Politics in Iran*, New Haven: Yale University Press.
Robertson, R. (1990) 'Mapping the global condition: globalization as the central concept', in M. Featherstone (ed.) *Global Culture*, London: Sage.
—— (1992) *Globalization: Social Theory and Global Culture*, London: Sage.
—— (1993) 'Community, society, globality and the category of religion', in E. Barker, J. A. Beckford and K. Dobbelaere (eds) *Secularization, Rationalism and Sectarianism*, Oxford: Clarendon.
Rostow, W. W. (1971) *Politics and the Stages of Growth*, Cambridge: Cambridge University Press.
Rowe, W. and Schelling, V. (1991) *Memory and Modernity: Popular Culture in Latin America*, London: Verso.
Roy, O. (1994) *The Failure of Political Islam*, London: I. B. Tauris.
Said, E. S. (1978) *Orientalism: Western Conceptions of the Orient*, London: Penguin.
Sakamoto, Y. (1995) 'Democratization, social movements and world order', in B. Hettne (ed.) *International Political Economy: Understanding Global Disorder*, Halifax: Fernwood.
Smith, C. (1991) *The Emergence of Liberation Theology: Radical Religion and Social Movement Theory*, Chicago: University of Chicago Press.
Sreberny-Mohammedi, A. and Mohammedi, A. (1994) *Small Media, Big Revolution*, Minneapolis: University of Minnesota Press.
Tarock, A. (1995) 'Fighting the enemy under a new banner', *Third World Quarterly* 16 (1).
Tripp, C. (1994) 'Sayyid Qutb: the political vision', in A. Rahnema, *Pioneers of Islamic Revival*, London: Zed.
Turner, B. (1994) *Orientalism, Postmodernism and Globalism*, London: Routledge.
Waters, M. (1995) *Globalization*, London: Routledge.
Zubaida, S. (1989) *Islam, the People and the State*, London: Routledge.

index